Technical Studies, tectonic explorations

*Notional considerations in developing a tectonic dissertation. This book is both warm-up and critical reading list for those studying architecture and about to embark on a Technical Study.*

Samson Adjei
Edited by Dr. Igea Troiani
Technical Editing by Ben Godber, Cíaran Malik

Technical Studies, tectonic explorations

First edition, published by East Winter Books
an imprint of Thought and Company Ltd,
13 Sunbury Workshops, Swanfield Street, London E2 7LF
www.eastwinter.com

While every effort has been made to check the accuracy and quality of the information given in this publication, neither the Author nor the Publisher accept any responsibility for the subsequent use of this information, for any errors or omissions that it may contain, or for any misunderstandings arising from it. However, as this text is a work in progress, please send corrections to: studio@eastwinter.com where every effort will be made to make amendments in a subsequent edition.

Book design and production by East Winter Books
Set in IBM Plex—with gDocs, CS5, and a 10yr old MacBook

This book is not intended as a substitute for the technical advice of Engineers. The reader should regularly consult an Engineer in matters relating to structural or environmental design and particularly with respect to any projects that may require private or public liability insurances. Some names and identifying details have been changed to protect the privacy of individuals.

© Samson Adjei and Dr. Igea Troiani. All rights reserved. No part of this book may be reprinted, reproduced or utilised in any form. The right of Samson Adjei to be identified as the author of this work has been asserted by him in accordance with sections 77 and 78 of the Copyright, Designs and Patents Act 1988. The publisher makes no representation, express or implied, with regard to the accuracy of the information contained in this book and cannot accept any legal responsibility or liability for any errors or omissions that may be made.

Except in the United States of America, this book is sold subject to the condition that it shall not, by way of trade or otherwise, be lent, re-sold, hired out or otherwise circulated without the publisher's prior consent in any form of binding or cover other than that in which it is published and without a similar condition including this condition being imposed on the subsequent purchaser.

ISBN: 978-0-9935046-2-4

**Acknowledgements**

I wish to acknowledge the kind assistance given by a number of people during the course of this project.

To begin with I would like to express my thanks to Eva Sopeoglou, Manja van de Worp, Sarah Moore, Maria Cheung and Paul Westmoreland along with Gregory Epps for their support, and critical motivation Aureliusz Kowalczyk, Dr. Ersi Loannidou, Stephen Baty, Dr. Savvas Verdis, Simon Gazzard, Avni Patel, Josef Huber, Jennifer Mensah, Patrick Weber and Sabine Storp, Simon Menges, Sam Chermayeff, Alison Crawshaw, Gort Scott, Erika Lanselle, Dyvik Kahlen, and the patience and curiosity of the RCA TS students. In addition the generous image and drawing donations of the many offices and creators featured throughout.
    Finally but foremost, sound judgement and no little skill has been exercised by Igea Troiani, Ben Godber and Cíaran Malik through their editorial surgery and general guidance.

# Image credits

Fig. 1_2  Heck, J. (1851). Iconographic Encyclopedia of Science Literature & Art. New York: Rudolph Garrigue, pp.Vol 2, Div. VII Architecture, Plate 22. Public Domain.

Fig. 1_3  Dinner Table, 2012. Photograph © Mauricio Alejo. www.mauricioalejo.com

Fig. 3_1  Diller + Scofidio Renfro. Blur Building, Expo.02 Switzerland 2002. By Norbert Aepli, Switzerland, CC BY 2.5,
https://commons.wikimedia.org/w/index.php?curid=796896

Fig. 3_2  Nimbus De.Groen 2017, © Berndnaut Smilde. Image courtesy of the artist and Ronchini Gallery

Fig. 3_3  Herzog & de Meuron, Laban Dance Center, 2002. By rucativava CC BY-SA 2.0,
https://commons.wikimedia.org/w/index.php?curid=1629193

Fig. 3_5  Phantom, A. (1874). Magic Lantern. London: Houlston & Sons, p.0. Public Domain.

Fig. 3_6  FEDEX Glass Works series, © Walead Beshty. Image courtesy the artist and Thomas Dane Gallery.

Fig. 3_8  The Fram. Public Domain.

Fig. 3_10–12  Tezuka Architects, Fuji Kindergarten, Japan. Images © Tezuka Architects

Fig. 3_13–15  Lycée français Alexandre Yersin, Hanoi, Vietnam. Images © Lacaton and Vassal

Fig. 3_16–17  Valerio Olgiati. EPFL Learning Center. Images © Archive Olgiati.

Fig. 3_18–21  Junya Ishigami. The Table. Images ©  junya.ishigami + associates.

Fig. 5_1  Kings College Chapel, 1446. Fan Vaulted ceiling. By TomAlt–Self-photographed, CC BY-SA 2.5,
https://commons.wikimedia.org/w/index.php?curid=2096776

| | |
|---|---|
| Fig. 6_1 | Frank Gehry, Guggenheim Museum, Bilbao. |
| | By Xauxa (Håkan Svensson) - Own work, CC BY-SA 3.0, |
| | https://commons.wikimedia.org/w/index.php?curid=17251189 |
| Fig. 6_2 | By Shustov - Own work, CC BY-SA 3.0, |
| | https://commons.wikimedia.org/w/index.php?curid=8934440 |
| Fig. 6_4 | Studio Mumbai Workshop. Photographs © Edmund Sumner |
| Fig. 6_5–7 | Venice 2016 Bienalle. Photographs © Biennale Atelier, |
| | Prof. Christian Kerez, ETH Zürich |
| Fig. 6_8 | Venice 2016 Bienalle. Photograph © Oliver Dubuis |
| | |
| Fig. 7_11–15 | Foerster-Baldenius and Raumlabor Architects, The Floating University, Berlin 2018. Photographs © Daniel Seiffert |
| Fig. 7_16,18 | The Eden Project, Cornwall. Drawings © Sir Nicholas Grimshaw and Partners. |
| Fig. 7_17 | The Eden Project, Cornwall. Photograph © Hufton + Crow |
| Fig. 7_19 | Axonometric Detail of the Crystal Palace build process. The Illustrated London News v.17 (1850.11.16). Public Domain. |
| Fig. 7_21 | London Zoo Snowdon Aviary. By heena_mistry, CC BY 2.0, |
| | https://commons.wikimedia.org/w/index.php?curid=9578136. |
| Fig. 7_22–28 | Prada Transformer/OMA. Images © OMA |
| Fig. 7_24 | Prada Transformer Image © Charlie Koolhaas; courtesy of AMO |
| Fig. 7_29 | Paul Smith Shop, Photograph © David Grandorge |
| Fig. 7_30–37 | Images © 6a Architects |
| Fig. 7_33 | Paul Smith Facade illustration. Image © Montrésor Partnership |

Note: The starting number of each figure position (Fig. #_...) indicates the chapter, the second relates the image.

Pictures and diagrams otherwise not listed above are by the author.
While every effort has been made to credit the present copyright holders, we apologise in advance for any unintentional omission or error, and will be pleased to insert the appropriate acknowledgement in any subsequent edition.

# Contents

1. **A Preamble** — 01
   Introduction, outlining: the relevance of TS; the aim of the book; and where it tries to sit within the wider context of construction publications.

2. **The Field (a reading list)** — 15
   A focussed categorised reading list, each with a brief description.

3. **Experimenta.** *Part I* — 25
   An informal discussion between two schools —one as a square the other as a circle.
   a. About mediating TS and Studio agendas
   b. About experimentation and prototyping

4. **Matter matters** — 55
   An essay posed as a list, a stream of occasionally dissociated points on a range of materials, aiming to reframe the materials of our environment.

5. **Experimenta.** *Part II* — 79
   A continued discussion between two schools.
   c. Reflection, reaction
   d. Case studies
   e. Document formation

6. **Approaching prototyping and the idea of the 1:1: a tangible component** — 91
   A look at the relevance of tangible things, and the roles they play in the design process.

7. **Next level** — 105
   For the more experienced TS student. Further explorations on the meanings of things, and the scope of detail composition.

# 1. A Preamble

## 1. A Preamble

*'Learn the rules like a pro, so you can break them like an artist'*
—Maxim, Anon, often attributed to P. Picasso

## TS

This book is aimed at the niche case of postgraduate architecture students engaging with their first Technical Study (TS)—where enquiry and learning outcomes are expected to be above those of undergraduate students collectively starting out. Whilst less prescriptive than a typical course guide, the core intentions of the book are to encourage a more exploratory attitude towards technical research, to suggest alternative ways of thinking about experimentation, and to provide a loose guide to compiling the Technical Study documentation for coursework submissions.

This text is not a 'how to' guide. There are already in-depth texts covering various processes in construction, structural design and environmental design from different perspectives in circulation. Instead, here, the core intent is to point towards a select few texts in the fields of Structural design, Environmental design, Prototyping and Construction. Additionally it should be taken as a introduction and loose discourse on an idea of the TS to help organise thoughts, and structure an approach towards building a substantial postgraduate level Technical Study. These recommended texts are typically written by experts in their fields, and in themselves tend to be more specifically focussed than is possible in the scope of this book. A handful of the texts and the information contained within them is addressed at practicing professionals. In this regard, your own judgement as a reader will need to be exercised (in conversation with a TS tutor and studio design tutors where possible) when considering the extent to which these texts might be applied to the constructs of a design project.

As such, this entire publication is in itself a form of preface to a TS and a foreword to the broader field of tectonics. For this reason chapter 2: 'The Field', stands as the most significant. Other chapters provide ways of approaching the TS and things to keep in mind to help shape a more engaging process and a more effective outcome. Various built example projects are interspersed throughout the book as demonstrations of alternative approaches towards tectonic solutions.

Though better described in Kenneth Frampton's Studies of Tectonic Culture, and the origins of interpretation through Gottfried Semper, this diagram *(Fig.1_1)* of word origins attempts to illustrate an etymology of 'tectonic' and 'stereotomic' along with their connections to processes of making. Where 'tectonic' is here crudely interpreted as the assembly of different materials in the making of form—structural materials plus lining, dressing, and weatherproofing materials, 'stereotomic' might be understood as the subtraction of material or the shaping of form e.g. the singular material as structural form and appearance, typically stone, or notionally Cross Laminated Timber (CLT). It might be possible to further attribute concrete, load bearing brick, and rammed earth to 'stereotomic'—even though concrete typically involves the additional material and process of reinforcement bars, and brick is combined with cement mortar.

Since others have defined these terms, for the purposes of this book use of 'tectonic' is imagined to incorporate the understandings of 'stereotomic', which is to say all forms of making with relation to buildings, whether principally additive or subtractive in approach. 'Tectonic' is also considered a subset of 'technical', where 'technical' has application in anything of sufficient complexity from microelectronics to baking. 'Tectonic' is specifically taken as the physical counterpoint to readings of architectural 'space'. As such, 'Technical Studies' can also be considered as 'stereo-tectonic explorations'.

The TS, often underrated and occasionally undervalued, is unlike other key forms of architectural coursework. This one document seeks to demonstrate technical competence as a designer and capability as a maker. There are many design disciplines that place an emphasis on portfolios; few of these explore polemic, occasionally provocative or often ambitious, approaches to reimagining our environments. Few to no other courses allow for the additional exploration of tectonic delivery (a TS) to the same extent. A deeply framed TS will add backbone to the most poetic of design proposals, but this isn't to say it does not have a poetry of its own.

As part of a greater tradition of practice, the TS aims to explore a path towards the inner workings of a project, this may or may not be the case for the main design thesis depending on the studio interests. The TS's success often lies in a balance between experimentation and rigour of approach, further conveyed through critical content such as descriptive and reflective thinking, and technical explorations of relevant case studies. However, it is also believed in most cases, if something is tedious to write (or make) then we should not be surprised if it is tedious to read (or experience). Conversely, if it's interesting to write (or make) it

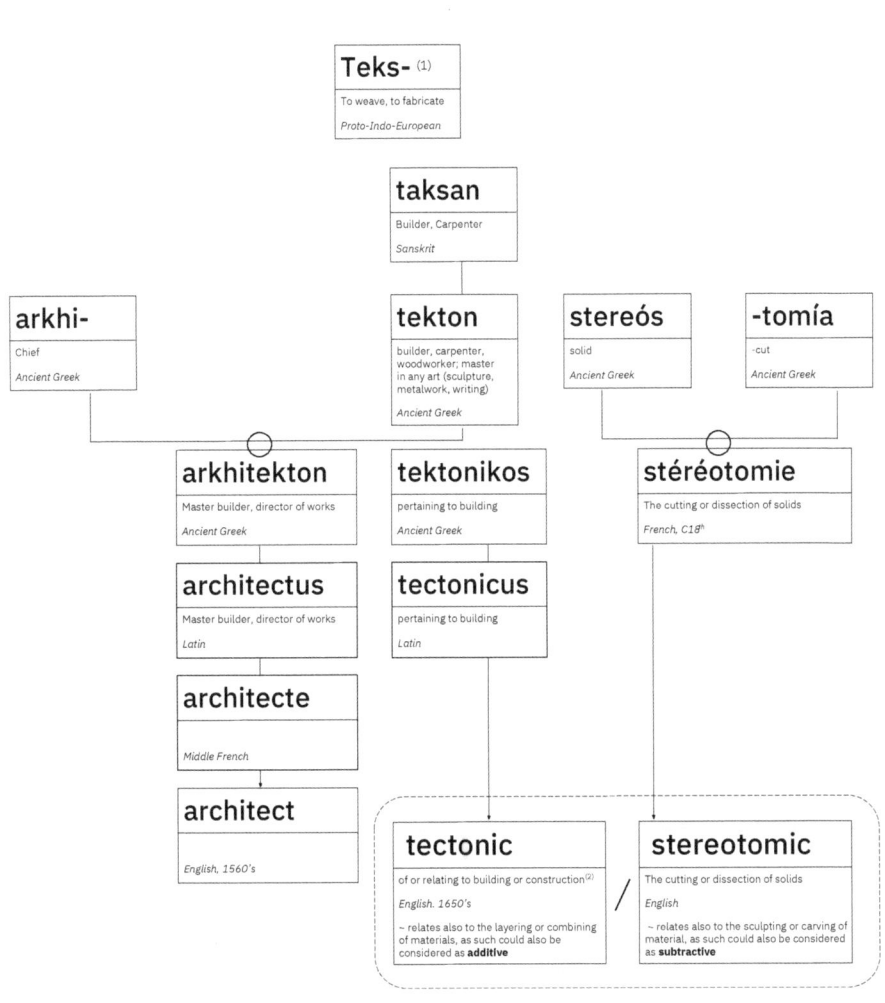

*fig.1_1: Tectonic and stereotomic etymologies*

fig.1_2: Heck, J. (1851). Iconographic Encyclopedia of Science Literature and Art. New York: Rudolph Garrigue, pp.Volume 2, Division VII Architecture, Plate 22

1. Tuscan capital and entablature
2. Doric pedestal and base, front elevation
3. Doric order upper column and entablature in the modillion style
4. Ionic order base
5. Ionic order capital and entablature
6. Corinthian pedestal and base
7. Corinthian cornice
8. Composite order pedestal and base
9. Composite capital and entablature

is often also, to some degree, more enjoyable to engage with. For this reason, it is understood that the TS direction and development is lead by the central design project. In turn, the central design project is guided by conversations with the design studio tutors and Technical tutors.

Unlike the standard dissertation format the subject areas of a TS are inherently more diverse and typically multidisciplinary, leading to numerous conversations and hopefully a broad understanding of the field. In professional practice this can lead towards more effective and informed conversations with consultants and construction team members.

A non-hierarchical list of common TS areas of research focus can consist of the following four categories:

- Structure
- Environment
- Prototyping
- Construction detailing

Case Studies and Experimentation are taken as present throughout, and Prototyping is considered in a little more depth through its own essay in Chapter 6 entitled 'Approaching Prototyping'. The remaining points, including formation of the document itself, are disassembled further below, and revisited again (for those reading with a little more experience) the final chapter 7 called 'Next Level'.

## *Technical explorations and room for reflection*

There is an advantage in approaching the compilation of the TS research and experimentation as an academic exercise. Moving beyond a simple description of the core project's working methods and those of various associated case-studies, the real value of a TS resides in demonstrating a balanced and critical understanding of differing approaches encountered in a design or design decision making process. This is especially interesting where such approaches stand in opposing positions. For example, in approaching a structural strategy, would it be better to use a steel or concrete frame? How might you assess these options given factors of environment, time, cost, longevity, and the context of a design project's requirements? These questions become more intriguing where differing strategies have generally opposing options. Critical evaluations —the careful weighing up of pros and cons of given option—should only be applied where relevant for a project i.e. only where they are interesting and useful to do, but preferably where one or both of the options in question are themselves in someway useful; where different pros or cons use otherwise unconventional arguments.

In this way a TS document might contain some five or more moments of critical reflection invoked through pondering options, but also self-critique in reviewing mistakes made, questions asked and new questions formed following experimentation. This is so for any good research paper. In addition there should be space for occasional thoughts on alternative issues that might come about if the project were to be constructed 'for real'.

## *Interpreting case studies*

This leads neatly to the role of the case study. The case study is an opportunity to apply critical understanding of another project's successes and failures. It is unusual for every part of a design project to be completely inventive or never before considered. So as to demonstrate that certain aspects of a scheme are novel or feasible in some way, it is useful to refer to previously built examples which have to different extents come close to what is being proposed.

In discussing a case study there are a few useful etiquette practices worth employing. The title, author/designer/artist/architect behind a case study must be referenced including a construction date, location, a brief description of the project—as in what it does and why or how it is relevant to this scheme. It is recommended to focus on key parts of the project, and produce new reinterpreted explanatory diagrams, to reflect or allude to the main strengths of the case study.

In addition to the array of detail drawings plans and project-related data to set context, a strong TS has a healthy mixture of various components. These include balanced reflection of critical approaches; engagement with physical explorations at 1:1; considered experimentation where appropriate; interestingly and widely referenced case studies; edited diagrams explaining key areas—all presented clearly, though, not necessarily in this order.

However, when starting out, it is important to not be overwhelmed by such criteria. It is often easier to momentarily place these aside, and start with a single outline diagram of the most interesting aspect of the main design thesis (as developing). One bite at a time. [1]

## Tectonics as text

The vast majority of architecture constructed before the 1990s was designed by hand and drawn up by hand. In all cases, numerous drawings at various scales would have been created manually with detail drawings repeatedly revised before and during construction.

A body of cultural knowledge and technical skills carried by the many architects and builders involved in a project is thus embedded in the collective work of a building. [2] From this perspective we might consider buildings simultaneously as books or even libraries of thought and technique and walls and elements as pages, junctions and details as words, components and decisions as letters. The more interesting task might be the idea of deciphering how to read and adopt the lessons of these tectonic texts.

This is worth pondering. As propositions, drawings (as a medium) matter but when translated through construction into physical matter they take on new demands: of use, and of weathering. Not all buildings stand up to these tests of time in the same way, and within any given project it is not uncommon to find a detail or two that hasn't performed as well as other details, or as intended. These small failures, and the clean successes can both be taken as critical lessons, in how not to and how to build. Even if the physical example's faults could be linked to workmanship errors we can still ask if these could have been reduced by more elegant and considered detailing, taking construction processes (sequences of assembly, also known as the 'build sequence') into consideration. What was the detail aiming to achieve? How has it failed? Would this same approach have succeeded elsewhere in the same building? What is it about its immediate context that differs? Here too the unsuccessful details deserve more attention and a little light analysis.

## Objectives in detailing

Details aim to get to the heart of or essence of an object's material condition and assemblage. These drawings address the methods needed to allow different

---

1. *'When eating an elephant, take one bite at a time'*, Creighton Williams Abrams Jr. (1914 –1974).
2. *'Cultural knowledge in architecture is embedded in the fine buildings, works of art and the fabric of the cities that exist around us. It is a matter of taking a careful look at these things to try to come to a deeper understanding about what the artists, architects and master builders of the past were trying to achieve.'* - Beigel, F. and Christou, P. (2014). *SAM Translations*. Basel: Christoph Merian, pp. 7.

materials to come together. Depending on the situation, details have many additional aims including: concealing screw fixings, maintaining water resistance, allowing ventilation, maintaining a line of insulation, connecting with a structure, coordinating with an adjacent detail to maintain a continuous surface, accommodating movement, etc. In the early drafts of a detail drawing it helps to have a clear idea, or a checklist perhaps, of the objectives it will be asked it to achieve.

The resulting schematic drawing will always need to demonstrate use of lineweights and clear annotation. Practising these throughout, not just at the finale, will lead towards a better outcome.

## *Structural design strategies*

A short summary list of thoughts to walk through on the way to thinking about the physical framework of a design project in question is to consider: adaptability, presence, mass, material, foundations, drainage, spans and the need for ...[some particular technique, or special process]. Some of these might not apply, and so might be omitted, leaving more room to reflect on the others. Keeping these in mind whilst a project develops will also help identify appropriately interesting case studies, which in turn provide useful lessons and strategies for adaptation.

Decisions regarding the core structural directions should be discussed with peers and studio tutors, and for the fundamental environmental strategy directions to follow. Where points overlap with existing course guidelines, then the course guidelines take priority.

## *Contemplating environmental design strategies*

The following should be considered a fallback where a guided approach to environmental design hasn't already been provided. As the core design project strategies unfold it is important to identify moments to focus on and explore in terms of spatial intentions. These intentions could be viewed as experiences, evaluated space by space, and adjusted for the project as a whole. In evaluating each of these spaces, perhaps three at most, we should aim to dissect and outline different environmental qualities of a given moment.

In turn, these qualities might be observed as thermal (mass, passive/ mechanical, varying, homogeneous), lighting (daylight, electric (various types)), ventilation (natural, pollution, heating), air (pressure, humidity, flow), acoustic, material integration, drainage and other layers as interconnected strategies. Each of these provides a framework within which to research and discuss attitudes and approaches. Ideally, this should be done lightly, before stepping back to help recognise common aims, and unifying strategies. In addressing the relevant layers it's important to address each as both an individual strategy and part of a whole entity including where appropriate, the structural strategy and the relationship it has with these environmental considerations. This addresses the interior, the environmental aspirations for within the project. Exterior considerations might address an alternate group of factors from drainage to energy and so forth.

Both inside and outside methods and approaches are better discussed in 'The Field's recommended reading. Whilst reviewing these texts thoughts developed from the above (or other) approaches should be used to select suitable case study investigations—just as in 'Structural design strategies' above.

## *The diagram*

Given the ability to reflect specific views of complex 3D scenarios, the propensity to create equally complex visually rich explanatory diagrams is understandable. But in another light, if a diagram is too difficult to understand then has it succeeded?

The more complex the subject of a diagram study the more restraint and editing of the graphic is needed to communicate. Avoid the temptation to make overly complicated diagrams of otherwise complex systems. Clearer simple diagrams require more time and editing than one might at first think but typically lead to better, more sophisticated, outcomes. Where it is difficult to say or show the things that need to be communicated, consider using variations of the same baseline diagram to express these different points.

Just as for the habitual ways of looking at case studies, it is worth developing a good 'eye' for other effective diagrams as observed in the wild. Addressing typography, looking at sums of colours and ranges, general clarity, annotation, and projection—perspective, axonometric, plan, section etc. Developing this background observational practice is a useful technique for enhancing future diagrams.

## *Referencing—creating connections*

Research itself exists within a continuous dialogue with past and future work by others with the greater aim of progressing understanding and ideas. Researchers are in some ways considered as temporary custodians of contemporary knowledge. Even if incomplete, the progress made towards resolving a problem may itself be found to make valuable contributions to other research projects at other points in time. This is reflected in the importance of referencing the acknowledgement of relevant previous research, and thus the act of positioning your own work within this continuum. Where research successes and failures are sufficiently interesting others in time may learn from the experiences of your own work and make references to it (citations) in their own developments. *'Citing a book from which you copied a sentence is paying a debt.'*[3] The nature of this ongoing conversation between studies allows us to make potentially significant progress by recognising and building on the earlier work of others, without repeatedly starting at zero.

## *Coherence—a form of weaving*

Though listed as separate subjects, the subheadings above coexist as enmeshed layers within the same design project, as one document. These chapters should ideally coordinate to present the feel of one thesis.

If, for a moment, we consider the TS thesis itself as a form of response or answer, then we can appreciate the guiding importance of framing a good question. Thus, recognising the core technical question(s) at the heart of your design interests and stating these early on will serve as a preface to the document—with respective sub questions opening relevant subsections Starting out with a clear question helps to anchor the study as a whole, providing a point of reference for subsequent explorations. Such a question should be of interest to the researcher and potentially others, researchable within the time frame and resources available, open-ended but focussed, neither too broad or too narrow, nor too objective or too subjective. The question(s) can be refined over the course of research and (crucially) should be formulated and discussed with your TS tutors.

Addressing chapter flow: at the beginning and end of each section it may help describe connections with the previous and upcoming respectively, thus easing the transitions, and forming a more fluid movement across the document as a whole.

Adjacent to this, consider how the visual cues provided by diagrams and images might be read in parallel with the core texts i.e. the idea of conveying a form of meta-narrative. Where diagrams are redrawn in the same illustrative language, accompanied by images/photography of equal framing and clarity, and consistently annotated to the same level then coherence become apparent.

However, any such coherence found in this book is coincidental. The following example projects reflect a small sample of available tectonic directions. There are many additional approaches to making which are not addressed here. Their absence is by no means considered any less interesting or relevant than those presented. Instead, a broader scope of examples are left to the greater depths explored in the recommended readings and beyond. Unlike the remaining text, your TS should benefit from a sharper agenda and a lighter touch. In many cases a central narrative built around a focused question, or a thesis, will provide a position from which to play with a spectrum of explorations.

In tandem with your Technical tutors, the reading list texts outlined in the following chapter will provide the knowledge and methods required for the technical resolution of your design projects.

*fig.1_3: Mauricio Alejo. Dinner Table, 2012*

3.  Eco, U., Mongiat Farina, C. and Farina, G. (2015). *How to write a thesis*. MIT Press; Translation edition (2015).

## 2. The Field

## 2. The Field

*There are dozens of useful textbooks discussing construction, and dozens more that specialise in materials and strategies. If there were just a few MUST-READ books for each subject you would recommend to all students starting out, which would they be and why?*

Fortunately there are already many excellent publications on construction and the many specialist subjects within the field. When approaching the range available it is worth remembering the majority are written with practitioners in mind. Others are for more standardised forms of project. A smaller group of books are written with the aim of introducing the variety of environmental, structural, and construction strategies available, and fewer still are designed for the purpose of assisting with a Technical Study. All and any of these texts will become important at differing stages of study and practice, some more so in different situations, and so there is an understanding that to try to rank titles by quality would be counterproductive.

To reduce bewilderment at the number of available titles, we have taken a moment to curate an opinionated selection as a study path and focussed way into the general subject of construction. This list is by no means exhaustive so it is well worth taking a moment to explore others during the process of refining your TS.

There is a distinction here between publications listed as effective resources [Resource] that contain a catalogue of useful information, and the more readable [Read] texts. These Resource books are useful reference documents—especially at the stages where a design thesis is just beginning to form or decisions need to be made between competing strategies. Resource books allow you to 'dip' into the relevant chapters for required information to build effective knowledge and improve the quality of a TS. It is important to consider these texts, and their relevant chapters, as useful starting points. These books provide a head start from which to follow up on footnotes and track down more specialist information.

Though here they are referred to as 'Resource' (reference books) you will often find the introductory chapters and subsequent chapter introductions are readable and therefore it is recommended some time is spent with each to understand it more fully.

For those described as 'Read' please aim to do so. Whether old or new these texts often overlap, in subject matter and approach. This should provide assurance whilst reinforcing an understanding of the common issues. In all cases, they will provide the insight and skills needed to recognise opportunities, build confidence, and improve design strategies. They will also help make better use of the Resource books.

**Overview**:

 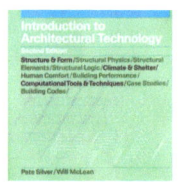 

This first publication almost serves as a road-map for the remaining texts and could be considered a must read, whilst the second overlaps and explores further perspectives. The suggested resource book will help to think about how to describe case studies and assist with factors to consider when detailing.

1. *Essential Read*: **How Buildings Work: The Natural Order of Architecture by Edward Allan** _ An easy introduction to various structural, construction, and environmental strategies that together encompass 'Building Physics'. There are numerous comprehensive diagrams to open up and illuminate each subject area. Although it doesn't fully touch on the poetic potential of tectonics, or possible implications of newer technologies, this is still a must-read for a holistic understanding of the matters at hand.

2. *Read:* **Introduction to Architectural Technology by William McLean, Peter Silver, and Dason Whitsett** _ Positioned between 'Read' and 'Resource' this text provides an encyclopedic map of design factors and explores emerging roles of computing, all backed by a concise array of short case studies.

3. *Resource*: **Introducing Architectural Tectonics: Exploring the Intersection of Design and Construction by Chad Schwartz** _ After an exploration of the meaning of 'tectonics' this book journeys through 20 different contemporary case-studies across eleven different countries to provide an overview of alternative structural and environmental design approaches. The graphics are clear and in many cases provide good examples of diagramming.

**Construction:**

In this section there is only one 'Reader' and one 'Advanced'. The latter is suitably dense, but worthwhile. The remaining Resource books also contain friendly introductions and specialised sections which should in time present their usefulness.

1. *Essential Read*: **The Architectural Detail by Edward R Ford** _ A highly readable and very thorough text discussing the language of detailing through an exploration of five different definitions.

2. *Advanced read*: **Studies in Tectonic Culture, The Poetics of Construction in Nineteenth and Twentieth Century Architecture by Kenneth Frampton** _ Provides a cultural and theoretical grounding for the modern history of tectonics. A solid foundational understanding for anyone serious about the 'craft' of construction. This one is not a quick read but the expanded depth of knowledge it addresses makes it worthwhile.

3. *Resource*: **Constructing Architecture by Andrea Deplazes** _ Compiled from ETH Zurich lecture notes this is a well presented anthology of subjects with effective diagrams and technical introductions. Be aware however that the details aren't always contextualised and so they won't always be suitable for your project's climate, detail scenario or situation.

4. *Resource*: **The Re-use Atlas by Duncan Baker-Brown** _ This text explores the global benefits of 'circular economy' thinking and provides for a more holistic contextual understanding of sustainable design agendas with regards to time, materials, and construction strategies. A diverse array of categorised case studies makes this a highly readable reference source.

5. *Resource*: **Architectural Detailing: Function, Constructability, Aesthetics by Edward Allen**_ A useful guide to detailing and the factors for consideration, with a myriad of descriptive illustrations. Aim to liaise with your Studio and Technical Studies tutors as your technical drawings develop.

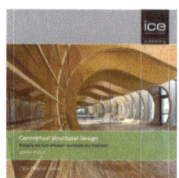  

**Structural Design:**

The trio of books on Structure are selected to cover the present past and possible future of structural design. In so doing, they also discuss its integral relationship with architecture, both pragmatically and theoretically.

1. *Essential Read*: **Conceptual Structural Design: Bridging the Gap Between Architects and Engineers by Olga Popovic Larsen** _ A short, readable, and easily digestible text with illustrations and (importantly) a variety of case studies.

2. *Read*: **Light, Wind, and Structure: The Mystery of the Master Builders by Robert Mark** _ A highly recommendable text, not too dissimilar in direction to **Developments in Structural Form by Rowland Mainstone**. This publication takes a more focussed interest in the medieval master builders and the ground breaking approaches to structural strategies they enacted through stone. There is still much to be learned from their adaptation of an otherwise prehistoric material.

3. *Read*: **Inside Smartgeometry: Expanding the Architectural Possibilities of Computational Design by Brady Peters & Terri Peters** _ An introduction to the ever expanding possibilities of computational modelling in Structural design. This approach is not to be confused with other more general digital exercises in form-making or generative design.

 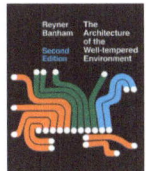  

**Environmental Design**:

With one listed resource book in this section all of the following books are recommended reads. These 'Read' texts are less focussed on the mechanics of 'environmental design' strategies, instead they more directly address the experience, methods, and motivations of this essential subject.

1. *Essential Read*: **Thermal Delight in Architecture by Lisa Heschong** _ A very short must-read text. It does for heat what **In Praise of Shadows by Junichiro Tanizaki** does for light and shade. Instead of thermally 'neutral' environments, Heschong makes a persuasive case for thermo-spatially variable environments where different ideas of 'warm' or 'cool' might relate to different spaces.

2. *Read*: **The Architecture of the Well-Tempered Environment by Reyner Banham** _ A classic text that narrates an alternative perspective on environmental design. By regarding *'modern architecture as a complete art of environmental control'* Banham *(1984: p.Jacket)* casts a different light on modernism, one that helps us to understand and engage with vernacular and contemporary design differently. A readable and highly recommended starting point for this discipline.

3. *Read:* **The Environmental Imagination by Dean Hawkes** _ Through research backed by first hand experience of the environmental character of selected projects, Hawkes' engaging study of design aspirations re positions the *technics* of environmental design within the *poetics* of architectural intention.

4. *Resource*: **101 RULES OF THUMB: For Sustainable Buildings and Cities by Huw Heywood** _ A different perspective. Addressing the greater environment, this friendly text is a comprehensive and highly digestible overview of sustainable design 'rules of thumb', bite sized micro lessons encapsulate a broad scope of technical and strategic points. To be applied in moderation.

**Prototyping and materials**:

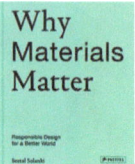   

These last four books address a little more closely the 1:1 prototyping aspect of a TS and may help in considerations of materiality and approaches towards making.

1. *Read*: **Why Materials Matter; Responsible Design for a Better World** by **Seetal Solanki** _ Described as a *'panoply of ideas, technologies, and creative efforts that focus on the earth's most basic elements, while also showing how these elements can be transformed into entirely new materials'* (S.Solanki 2018). Especially recommended for those making material choices or engaged in prototyping. Few of the examples target the architectural scale, which leaves room for creative re-appropriation or combination in the task of imagining architectural possibilities.

2. *Read*: **Stuff Matters: The Strange Stories of the Marvellous Materials that Shape Our Man-made World** by **Mark Miodownik** _ This illuminating book is an informal yet informative perspective of the materials we use, from steel to chocolate. Here Miodownik guides us on a fascinating journey of material histories, properties, cultural values, compositions and innovations.

3. *Resource*: **Construction Materials Manual** by **Hegger Auch-Schwelk and Fuchs Rosenkranz** _ Part of the informative '***DETAIL***' series of publications.

4. *Read*: **[eBook] Prototyping Architecture** by **Michael Stacey** _ Useful examples of 1:1 making and the role prototyping plays as a form of experimentation in practice. One of few books of this nature. The approaches described here may help to inform your own when considering the role of this exercise in your TS.

## The Field: summary bibliography (Referencing):

**Overview:**

1. Allen, E. (2005). *How buildings work*. Oxford: Oxford University Press.
2. McLean, W., Silver, P. and Whitsett, D. (2013). *Introduction to Architectural Technology*. London: Laurence King.
3. Schwartz, C. and Ford, E. (2016). *Introducing architectural tectonics*. Birkshire: Routledge.

**Construction:**

1. Ford, E. (2011). *The Architectural Detail*. Princeton Architectural Press
2. Frampton, K. and Cava, J. (2007). *Studies in tectonic culture*. Chicago, IL: Graham Foundation.
3. Deplazes, A. (2008). *Constructing architecture*. Basel: Birkhäuser.
4. Baker-Brown, D. (2017). *The Re-use Atlas*. London: RIBA Publishing.
5. Allen, E. and Rand, P. (2007). *Architectural detailing*. New York: John Wiley & Sons.

**Structural Design:**

1. Larsen, O. and Tyas, A. (2003). *Conceptual structural design*. London: Thomas Telford.
2. Mark, R. (1994). *Light, wind, and structure : the mystery of the master builders*. London: MIT Press.
3. Peters, B. and Peters, T. (2013). *Inside Smartgeometry*. Chichester: Wiley.

**Environmental Design:**

1. Heschong, L. (1978). *Thermal delight in architecture*. MIT Press.
1. Tanizaki, J., Harper, T. and Seidensticker, E. (2009). *In praise of shadows*. Vancouver: Crane Library at UBC.
2. Banham, R. (1984). *The Architecture of the Well-Tempered Environment*. University Of Chicago Press.
3. Hawkes, D. (2014). *The Environmental imagination*. 2nd ed. New York: Routledge.
4. Heywood, H. (2015). *101 RULES OF THUMB: For Sustainable Buildings and Cities*. Newcastle: RIBA Publishing.

**Prototyping and Materials:**

1. Solanki, S. and Corbin, L. (2018). *Why materials matter. Responsible design for a better world*. München: Prestel.
2. Miodownik, M. (2014). *Stuff Matters*. London: Penguin Books.
3. Rosenkranz, T., Auch-Schwelk, V., Fuchs, M. and Hegger, M. (2013). *Construction Materials Manual*. Basel: De Gruyter.
4. Stacey, M. (2013). *Prototyping architecture*. Riverside Architectural Press.

*'Some books are to be tasted, others to be swallowed, and some few to be chewed and digested: that is, some books are to be read only in parts, others to be read, but not curiously, and some few to be read wholly, and with diligence and attention'* - Sir Francis Bacon, Essays (1625), 'Of Studies'.

Managing just one or the first of each section should help elevate the quality and depth of your eventual TS and, more importantly, your understanding of the field. Keep in mind the interests and objectives driving your design thesis as a means to help navigate the array of options some of these texts may present. They present new opportunities to explore new ideas and learning in future projects. As such perhaps try to avoid placing too much emphasis on one over another, and where new ideas or directions make an appeal, be sure to discuss with colleagues, design tutors and technical tutors as notions develop.

## 3. Experimenta. Part I:
### The Circle and the Square.
*Monday*

## 3. *Experimenta. Part I*
*The Circle and the Square: Monday*

## *Mediating TS and a studio agenda.*
*Scene: At a table, a half full coffee cup, a reopened flask. The table is in the Serpentine pavilion designed by Frida Escobedo. Warm air, overcast sky. Two Technical Study tutors discuss.*

Circle:
    There are many situations where studio interests operate at scales not obviously compatible with the format of a Technical Study (TS), such as urban, landscape or furniture related projects.
    This is in principle where TS are led by a design studio's agenda. How would a student best position the physical nature of a project to work tectonically whilst maintaining a studio's process?

Square:
    A question of mediation, I feel typically, the TS or at least the idea of it is ultimately preoccupied with the tangible, something of enough complexity to form engaging conversations, reflections and experiments.

◯   What about those whose design proposals are entirely intangible? I could imagine, for example, if the student was skilled or confident enough it might even be possible to attempt a TS on an artificial-real cloud.

▢   How might that translate?

◯   So, you could argue; as a recognisable entity a cloud embodies a certain weight of water, volume. It's part of its own family of variations; typology, exists at a height with specific internal forces governing formation; structure?, it's entirely a part of its environment and as such subject to changes in air pressure, thermal differences and so on. Different inherent cloud structures can exhibit fascinating interplays with light diffusion. Case studies could include Diller and Scofidio's 2002 *Blur Building* *(Fig.3_1)* or Berndnaut Smilde's *Nimbus Project* *(Fig.3_2)* to name a few. There are many more interesting exercises that could be drawn out.

☐ Amusing. An interesting hypothesis, but these examples relate to controllable scenarios, surely. Diller and Scofidio's [Blur building] is, at least in the sense of a 'design', and though we have interesting ground level man made approximations of mist-making-structures the idea of inhabitation was only intended as a passing experience, momentary occupation of an art installation.

I would imagine subsequent issues of inhabitation and control might then be explored in alternative case studies. I like and appreciate the premise of it all. Although, I might worry this, approach risks being considered a novelty, where the concept overpowers the possibility of greater depths [...].

○ And yet this is the foundation of so many interesting discoveries. The start of a more involved process of experimentation. No?

☐ I wouldn't disagree with that in principle, it's just that sometimes when the drive to be novel dominates the potential scope of a TS [the TS] tends to narrow, and outcomes tend to lack depth. Or to frame it another way; in this example, without similar efforts in tackling the constraints and opportunities of 'comfort', [that is] some attempt of inhabitation, some dimension of detail drawing and so forth, the central invention—the cloud—becomes dismissable.

To be unconventional still means to be in some part conventional. If the sole aim is to work outside the rules then there are, I believe, many perspectives and useful lessons we risk overlooking

○ So you think all projects should be roof and wall structures?

*fig.3_1: Diller + Scofidio Renfro.*
*Blur Building, Expo.02 Switzerland 2002*

☐ Well, no. ...That sounds crude. I have to concede I've seen TS's for projects barely larger in scale than street furniture, and others essentially [are] urban landscaping lacking 'traditional' structures though still managing to impress as thorough pieces of technical research. So yes, there are exceptions. In your cloud example the research would be found in any resource on meteorology. Experimentation could be interesting but due to the one material there would be little to detail—if anything at all, thus what would be left to deliver? Though I shouldn't be saying it, I've had similar questions for ice based building projects in the past.

It's a balancing task of achieving an engaging but not impossible degree of complexity, and in this regard the TS agenda, and objectives, are often the more straightforward. Before this there is of course the studio's own overarching agenda, and (I would argue) before both of these there should exist the student's own agenda. There may need to be some mediation between all factors in leaning towards the best outcome.

◯ This to me sounds reasonable, but in addition to the TS tutors I might add that studio tutors are a source of support for TS, and so the distance between these positions should reduce through technical and studio communication. It's common for projects to wander and find themselves in a bind, attempting to pull together drastically different viewpoints where a simple [cross tech-studio] conversation would help to identify more straightforward narratives.

Nimbus Project, De Groen

In the Nimbus Project, Berndnaut Smilde creates interior cloud artworks in a variety of public and gallery environments. For this to work the air must be still (no airflow) damp and cold. The nature of clouds created changes in relation to these parameters. Given this environment water vapour is then sprayed in by hand, after which a smoke machine is used to send a short burst of 'smoke' into the recently vaporised space. For photography Smilde produces around a hundred clouds, each lasting approximately ten seconds, in order to achieve the right concentration and texture.

*fig.3_2: Berndnaut Smilde. Nimbus Project, 2017*

## *About experimentation and prototyping*
*Scene: Walking through the ceramics workshops on route to the computer lab, book in hand. Dusty air, like the flour haze of a bakery. Grey hues, hard reverberating surfaces.*

○ On discussing forms of experimentation, we briefly mentioned the idea of a cloud as [an] architectural notion and basis of a TS. I'd like to think about ground level forms of experimentation, those more readily relatable to practice, and TS's.

☐ So for example prototyping? There is a long and interesting tradition of this in practice, dating back to antiquity where ambitious forms of stone or timber structures were often empirically tested and new structural proportions explored through partial mock-ups. Later the fruits of form making experiments with examples such as Auguste Perret and Henri Sauvage's developments in reinforced concrete and Charles and Ray Eames' work on steamed plywood and fibreglass. Yet now the idea of [architectural] offices taking time out to trial new methods is still somehow exceptional. I find this interesting. Though in fairness, and in practice there are difficulties coordinating the time with other office processes and economic demands, and persuading a client that something 'untried' might be conducted at their expense is sometimes a problem. Or it's done at the offices expense, and either way—if successful—often needs to be followed up by various fire, maintenance, durability, health and safety tests before production.

Let's not forget the darker legacy of building products such as asbestos, which at some point in the mid 19th Century was taken up as a good idea.

○ An argument against innovation?

☐ No it's actually one of the more engaging aspects of practice, sorry that didn't start off so well. There are of course exceptions where offices who value research are keen to invest, and others where there are strong client relationships contributing to industry leading projects. However, while studying, the index of opportunity to try new things and 'see what happens' is much greater *(Fig.3_5)*. This is where I was attempting to go.

It's an interesting challenge to design a new form of experiment, or even to cobble together approaches from previous explorations, and then to spot opportunities, possibly buried deep within results. In most cases, where the results aren't what you were hoping for, it's best to use outputs as the material input of the next part of the experiment, to iterate feedback until more interesting outcomes are surfaced.

○ In other words, 'if the result isn't great, then the experiment isn't over'. There is a risk of going to too far here. Things could easily start to look desperate.

☐ Well, the key word here is risk, obviously not in the sense of anyone's health, but in the sense that sometimes results don't always look convincing, or aren't always useable, but even in those instances there's still much to document about the process, namely the intentions and design ambitions.

This, I should say, and I hope we can agree, is the basis of a real experiment. If we know 100% what the outcome will look like, then was it an experiment? No, I think it is more like a rehearsal. There is a quote from Evgeny Morozov addressing this position:

*'Creative experimentation propels our culture forward. That our stories of innovation tend to glorify the breakthroughs and edit out all the experimental mistakes doesn't mean that mistakes play a trivial role. As any artist or scientist knows, without some protected, even sacred space for mistakes, innovation would cease.'* [1]

But I think another key notion, one that takes some experience or to relate, is 'judgement'.

○ I prefer a shorter alternative, by Prof. Neville Scarfe but often incorrectly attributed to Albert Einstein:

*'The highest form of research is essentially play'* [2]

[...] In my experience the starting conditions, the parameters of play, and parameters at play, are a critical factor in determining intelligible outcomes. As you've mentioned already, 'judgement' is needed to achieve that balance of familiar and unknown, allowing for intuition and experience, to help take aim at something new.

You mentioned somthing about designing 'new forms of experiment'?

☐ […] How you go about conducting an experiment forms the bulk of the experience. Designing the experiment itself, like cooking, involves breaking down a process into component steps as needed. Not thinking of the outcome as the product of a singular action […]. To use a cooking analogy, that would be akin to peeling a potato. Further steps are needed to achieve something of interest. And so here we can explore the idea of baking a sample before then sealing it and sandwiching it, then pressing it, perforating it, dipping it and so forth.

To see what happens.

I believe an experiment can sometimes be as simple as placing two things together, and reassessing, a reductive approach. There can be a simple pleasure in trial and error, especially where new processes or material combinations are involved. It's important to not force an outcome whilst allowing oneself space to reflect on the results, and [have] time to redo parts when inspiration hits.

So in professional practice, where projects are already securely grounded, you can see that many designers jump at the opportunity to engage in experimentation because it's a break from normal processes. […] Students really should make the most of the chance while able, whether it's a facade element, a hinge, a piece of ceiling, a new type of door or whole new philosophy of building element. In amongst the many studio events, it's largely just a question of recognising and creating room for these moments to occur. Otherwise, watching that time pass by might cause frustration.

○ All but still mediated with the studio's agenda and the project's own design path… Depending on the studio schedules, it might also be important to consider when the earliest, and the latest opportunities to carry out an experiment might be.

☐ In the past I've noticed frustration occurs where […] exercises are seen as a one-off tasks, when, the stronger projects find more than one moment to attempt smaller experiments, more like 'idea tests'. Sometimes this leads to bigger ideas for tests. Sometimes they're sequential; sometimes different. Either way, the diversity of thoughts to explore and things to discuss in the TS is much broader, on top of which, trying out different ideas over a period, […] tends to be less stressful than building up to the one large experiment and not leaving enough time to react.

1. Morozov, E. (2018). *'Is Smart Making Us Dumb?'* [online] WSJ. Available at: https://www.wsj.com/
2. The journal *"Childhood Education"* published an article titled *"Play is Education"* by N. V. Scarfe, 1962.

○ We do also like to see things concluded, as in a written verdict and reflection, even if the physical results aren't yet there, though it should hopefully go without saying.

☐ I've often employed a simple mini template for experiments, based around four headings: Intention, Method, Apparatus (Equipment), Result. *(Fig.3_4)* Or: Planning, Prediction, Variables, Fair Test. They're just optional frameworks around which to position notes and records, and to help with developing an experiment.

Abstractly, I'm also reminded of 'improv' and film-making, impromptu props and multiple takes. Or even Jazz scores where particular phrases are left blank to accommodate improvisation, though this doesn't however mean performers are locked into other notes. Not the most direct of analogies but...

○ No, but it's sort of appreciated, and speaking of film there's a form of example in Lars Von Trier's *The Five Obstructions*.[3] Being one of his films there are predictably distasteful moments which aren't relevant to this discussion, but despite this it has a thesis that speaks of creativity and constraints. In it Von Trier invites Jorgen Leth, his former film school tutor, to recreate his 1967 piece titled *The Perfect Human*.[4] The premise is, with each recreation an additional 'obstruction' would be added to the brief, forcing a significant change to each subsequent iteration. The obstructions are themselves improvised with each recording and so Leth is forced to redo his work with accumulating restrictions. Anyway, the point being on the whole, with exceptions, the results build with each outdoing the previous.

This, in another sense, speaks to that experience of having to redo work; those situations I'm sure we have all had, where a computer crashes, or a mistake is made and unsaved work is lost or written over. Resigned to the prospect of wasting a block of time, we find the only way forward is to start again, or almost. Somehow, in less time than we thought the work is done, and somehow the replacement work is fresher and enhanced. The initial aim is to reproduce the work, but the outcome itself turns out to be an iteration. In reality this is not always the case, but as an experience it's assuring and even magical when it works out. To return to the here, it is also true of work in practice, every time a revision is made due to a change in brief or other criteria, except of course where costs are cut and dreams are quashed.

But I've digressed a little from experimentation as this is mostly speaking of change and adaptation. In TS there's often the ambition to innovate, whether it's a whole new form of experiment, or a whole new form of result. In order to neutralise this pressure, I wonder if there's a way of progressing a good TS in the absence of an 'invention' per se?

☐ At some point in the introduction of Kenneth Frampton's Studies in Tectonic Culture[5], the sense of this relationship between tradition and innovation is discussed. Frampton quotes Alvaro Siza who explains; *'architects don't invent anything, they transform reality'*. In this way, I believe, there is scope in alteration that can be equal to innovation. For every comparatively novel facade treatment by Herzog and De Meuron *(Fig.3_3)*, there is an interesting spatial exploration by Valerio Olgiati, and yet practitioners like these develop their relationship with materials not just to invent, but to understand what the material or the detail does and doesn't 'want' to be; refining an integral relationship between detail and spatial expression.

○ Indeed, the 1:1 prototype is typically predetermined to become an artefact through a focus on material and the physical.

The 1:1 might be an object, but as we are also concerned with spatial practices it might even be an augmentation of a space, to recreate a specific intention of atmosphere, microclimate, or other form of experience with a presence in a design project. This type of experimentation might benefit from techniques otherwise familiar to spatially engaging art installations, or even methods used by stage illusionists. The result is of a manipulation of temperature, airflow, lighting, acoustics and optics (the gaze, the view) to bring to mind a precise sense of a key moment (or moments) embedded as part of a design proposal. These are technical manipulations of an atmosphere or spatial experience.

*fig.3_3: Herzog & de Meuron, Laban Dance Center, facade.*
*A play betwen transpaerent (glass) and transluscent (polycarbonate) materials*

☐ A useful perspective, though a little less portable than an object. A change of frame from object-artefact to environment-experience. One for consideration, especially in regards to notions of learning from other forms of practice. This brings to mind the positions of those who have argued for something like an intra-architectonic approach; opposed to reaching outside the discipline of architecture itself for references or inspiration. This appears to be connected with a growing confidence in the creative potential of architectonics.

◯ Could this be a problem for the 'new'? To use another abstract analogy, in almost every past and present iteration of Sir Arthur Conan Doyle's *Sherlock Holmes*, the eponymous hero has an ongoing process of conducting seemingly random but creative experiments. This form of constant practice creates a foundation of knowledge which then translates to creative problem solving. 'Newness' and translation are the means through which Holmes is able to keep ahead of his foes own inventions. The majority of these experiments draw from a spectrum of other disciplines, spanning from music to mathematics.

☐ Yet we are responding to changes in climate and attitudes to energy consumption, even innovations such as Passivhaus *(Fig.3_7)*, itself a result of vernacular studies and prototyping, are continually updating. Do you feel there may not be enough tectonic scope within architecture itself to move forward?

◯ Not at all, it's just that few to no other disciplines have to date developed their individual disciplinary positions in isolation. But [...] if we can look around with a creative curiosity, we might start to recognise common grounds for exploration in other disciplines. Be it theatre *(Fig.3_5)*, film, sculpture, coding, or even naval engineering *(Fig.3_8–9)*, there are elements of spatial practice which for one reason or another have been further explored through parallel practices, sometimes in ways architecture hasn't yet had the same opportunity to do so. I'm thinking of the spatial powers attributed to common props in the stage projects of Simon Mc Burnley's *Theatre Complicité*, or the optical and geometric explorations of artists such as Olafur Eliasson, or even the more advanced engineering developments attributed to the International Space Station spring to mind. If there are ways of thinking, leading to the enrichment of architecture, then is this a bad thing? Architects didn't invent glass or steel but the benefits of technology transfer go beyond measure.

Conversely, though I appreciate it, I'm not sure I see the full potential of an introspective, or even intra-architectonic as you put it, stance to fully discuss or engage with the changing landscapes of human activity?

Pepper's Ghost

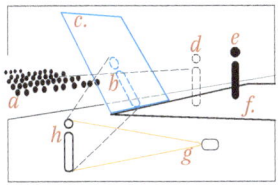

a. Audience
b. Reflected image
c. Glass
d. Perceived reflected image
e. Stage
f. Real stage context
g. 'Magic Lantern (Light source)
h. 'Pepper's Ghost' source image

This form of stage illusion takes advantage of the simultaneous transparent and reflective properties of glass to combine the view of a background with a projected source image. Its success depends the careful balance of light and dark on three sides: the audience (darkest), the stage, and the illusion subject (dark backdrops, controlled spot lighting). Ambient and stray lighting is minimised to avoid revealing the glass itself.

fig.3_5: The holographic theatrical stage effect of 'Pepper's Ghost'[6]

☐ Applicable points, but then as you like to look to others I will use the analogy of language and remind us both that while this language has a base set of twenty-six letters and nine numbers it's worth reflecting on the number of 'new' conversations we are still able to have. I think this in someway speaks for the idea of becoming more familiar with the vocabulary of materials and methods at hand, and thus maybe even more eloquent in the communication of architecture, expressing more than simply technique and moving beyond novelty. Even today practitioners are continuing to develop thesis in structural and environmental approaches. The spatial-environmental strategies of Tezuka Architects *(Fig.3_10–12)* and Lacaton and Vassal *(Fig.3_13–15)*, or the structural studies of offices such as Valerio Olgiati *(Fig.3_16–17)* and in a different way Junya Ishigami *(Fig.3_18–21)*, are just a few that spring to mind. This is the challenge, of how to imagine a work able to define its own position; perhaps between our viewpoints? Or the creative interpretation of other approaches, balanced with an internal meditation of its own.

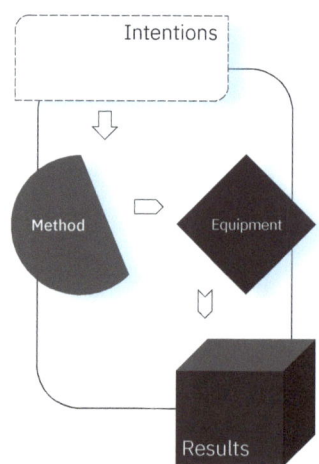

*fig.3_5: An example process for describing an experiment*

3. *The Five Obstructions*. (2005). [DVD] Directed by L. Von Trier and J. Leth. Denmark: Panic Productions.
4. *The Perfect Human*. (1967). [film] Directed by J. Leth. Denmark.
5. Frampton, K. and Cava, J. (2007). **Studies in tectonic culture**. Chicago, IL: Graham Foundation. Ch.1, pp. 25.
6. Phantom, A. (1874). ***Magic Lantern, How to buy and how to use it also how to raise a ghost***. London: Houlston & Sons, p.0.

## The FEDEX Glass Works Series

Interpreted here as embodying a notional form of experimentation—and part of a series spanning 2005–2014—artist Walead Beshty's FEDEX project absorbs the wear and tear encountered through transit, and poor handling.
The works are created simply as unbroken glass boxes, packaged and sent across the country (US) to various galleries, the resulting works are exhibited as received. Each piece is considered a 'fingerprint' document of the journey endured. As a form of experimentation the 'success' factor of the work is relayed through material wear and handling. Both image and title reflect one particular instance of the work during its lifetime.

fig.3_6: *Walead Beshty. FedEx® Large Box Priority Overnight Los Angeles-Berlin trk#857875945480, Berlin-Los Angeles trk#860752212570 , 2007*

## Passivhaus, an introduction

Passivhaus (Passive House) originated from a conversation between Bo Adamson (Sweden) and Prof Wolfgang Feist (Germany) in 1988, and drawing on observations of older traditional vernacular buildings as far a field as Iran and China.

From this discussion the idea of passive heating and cooling developed through a series of research projects into a set of standards and the formation of the Passivhaus Institute in Germany. This methodology embodies a rigorous approach to insulating and ventilating the building envelope, whilst offering substantial savings on energy consumption — as compared to common environmental design practices. Originally suited for northerly cold climates it has since been successfully adapted to many climate extremes.

A key focus is on the retention of heat through optimum use of solar gain to heat an environment, backed by rigorous avoidance of uncontrolled air leakage or drafts through the building envelope, and consistent use of insulation to keep heat out and/or in as required. In principle interior thermal comfort is provided by a combination of body heat and daylight, with only a minimal amount of additional heating or cooling required during seasonal extremes. Warmth from expelled air is recovered by transfefring it back to the fresh filtered incoming air via a Heat Exchange Unit.

Air quality and comfort underline the core strategies: 'A Passivhaus is a building in which thermal comfort can be achieved solely by post-heating or post-cooling the fresh air flow required for a good indoor air quality, without the need for additional recirculation of air.' - Passivhaus Institut (PHI)

As this method typically uses ducts for air and heat distribution, an understanding of air ducting systems is useful when detailing to resolve potential issues of unwanted sound transmission between rooms and the spread of smoke (fire).

Further research is encouraged as the Passivhaus strategy allows for other and more interesting possibilities than expressed in this brief introduction. In doing so it is worth revisiting the opening quote of chapter 1: A Preamble.

Building Envelope:
*a.* Rainscreen/weatherproof line
*b.* High performance insulation (continuous line) Uw=< 0.15 W/m²K
*c.* Openable low-e (low emissivity) triple glazed window. Low-e glass coatings reflect long-wave infrared thermal energy to reduce heat loss during colder months. Conversely they can also reduce solar gain during the summer months, to reduce air conditioning demands. Uw<= 0.8 W/m²K.
*d.* Air tightness barrier (continuous line) prevents airborne moisture vapour (including heat energy) of an interior from entering the building fabric where it can cause mould, rot or decay. Alternative 'Breathing wall' approaches forgo a vapour barrier and ensure vapour (moisture) is transmitted across the building envelope to the exterior whilst retaining heat energy. This is method is better defined as 'Vapour permeable' construction— where only air moisture is pushed through and out, whilst warm airflow is retained.
*e.* Primary structure. Optionally thermally massive.
*f.* Solar assistance, of which there are several variations and sub-systems available. Ranging from Photo-Voltaic (PV) electricity generating to direct water heating systems. Resulting energy may be used to drive a heat-exchange pump system'.

Interior:
*g.* Cold external air intake filter
*h.* Cold stale air exhaust
*i.* Heat recovery ventilation unit
*j.* Warm stale air extracted from bathrooms/kitchens.
*k.* Warmed fresh air sent to bedrooms and living areas
*l.* [Optional] ground based heat pump. Typically 15–150m depth (UK). Using the thermal mass of the ground, as heated through solar gain, to assist in the warming of external cold air.

Heat Recovery:
*m.* Filtered cold air input
*n.* Thermal heat transfer. Heat from the warm stale are is conducted to warm the cold external air, without mixing either air channels. For Passivhaus systems >75% of the heat energy from warm stale air is exchanged to heat the incoming fresh air.
*o.* Mechanical air pump (active) system. Optionally non mechanical (passive) systems are possible but offer reduced heat energy transfer percentages of <60% typically.

*fig.3_7: [Opposite] Passivehaus strategy*

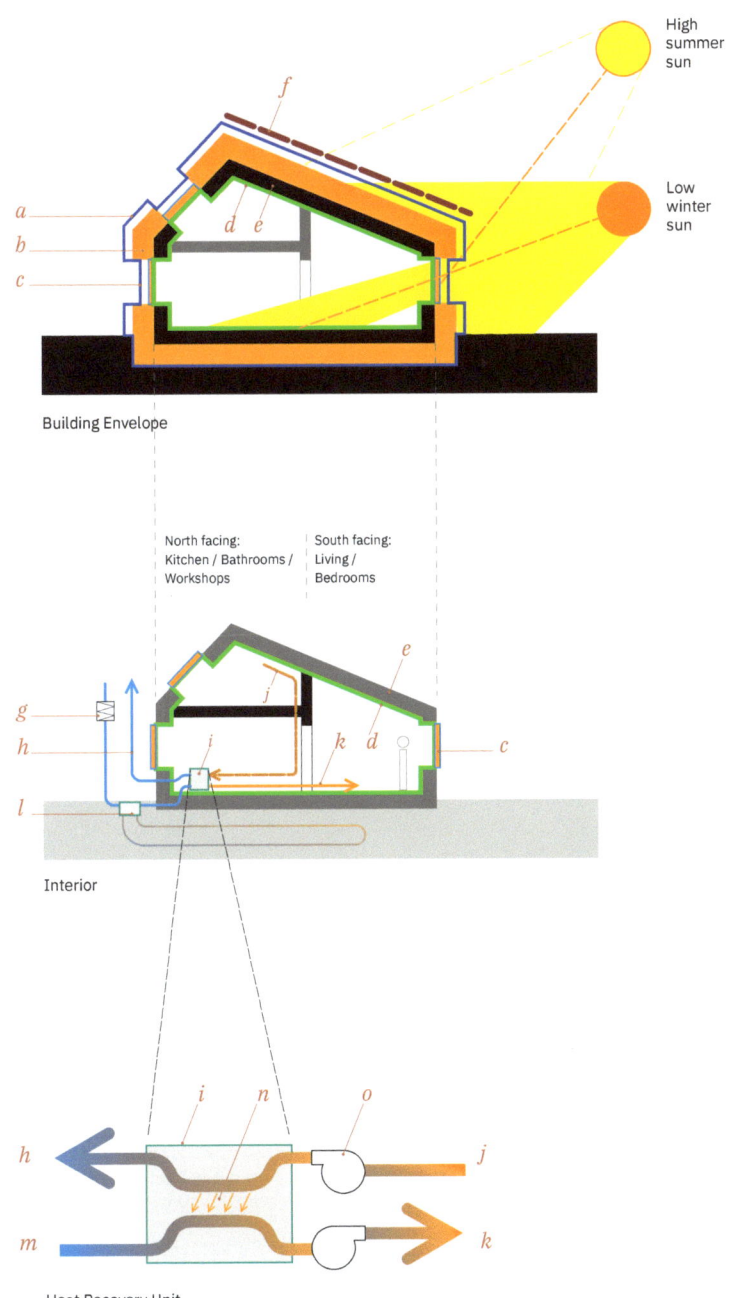

## The Fram, Arctic Ocean

Unknown to a few, but credited by some[7] as 'the first fully functioning Passive House' was in fact an exploration ship designed and lead by Fridtjof Nansen in 1893 and built to reach the North Pole with a crew of twelve. Though it didn't complete its original mission, it accomplished a number of geographical research achievements during its three year journey—where its crew would spend up to eight months at a time in its compact comfortable highly insulated living quarters before eventually returning all crew safely.

The layering of different materials with functions assigned to each, and its successful mediation of adversarial weather conditions, placed the Fram's environmental design approach significantly far ahead of it's time.

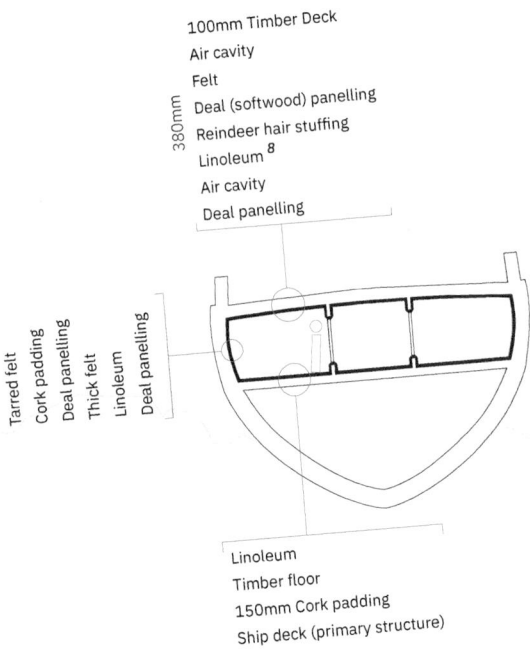

As Nansen notes through his journal[9] of the voyage:

'The skylight which was most exposed to the cold was protected by three panes of glass one within the other, and in various other ways. One of the greatest difficulties of life on board ship which former Arctic expeditions had had to contend with, was that moisture collecting on the cold outside walls either froze at once or ran down in streams into the berths and on to the floor. Thus it was not unusual to find the mattresses converted into more or less solid masses of ice. We, however, by these arrangements, entirely avoided such an unpleasant state of things, and when the fire was lighted in the saloon there was not a trace of moisture on the walls even in the sleeping cabins. ...

The Fram is a warm, cosy abode. Whether the thermometer stands at 22°C above zero or at 22°C below it, we have no fire in the stove. The ventilation is excellent, especially since we rigged up the air sail, which sends a whole winter's cold in through the ventilator; yet in spite of this we sit here warm and comfortable, with only a lamp burning. I am thinking of having the stove removed altogether; it is only in the way. At least as far as our protection from the winter cold is concerned, my calculations have turned out well.'

*fig.3_8: [Opposite] Simplified list of materials as surmised from Nansen's journal. Material sequence unordered*

*fig.3_9: [Above] The Fram. circa.1895*

7. Passipedia.org. (2019). *The Passive House – historical review [ ]*. [online] Available at: https://passipedia.org/basics/the_passive_house_-_historical_review [Accessed 10 Jan. 2019].
8. Linoleum: a hard but flexible material similar in feel to vinyl, commonly composed of solidified linseed oil (linoxyn), pine rosin, ground cork dust, wood flour, and mineral fillers such as calcium carbonate, most commonly applied to a burlap or canvas backing.
9. Nansen, F. (2008). *Farthest north.* New York: Skyhorse Pub., pp.44–45, 153.

# Fuji Kindergarten, Japan

Foregoing local requirements for the installation of mechanical air conditioning systems Tezuka Architects Kindergarten in Fuji's Tachikawa district, Japan, has a closer relationship to its microclimate than many others, resulting in a learning environment that is healthy by design.

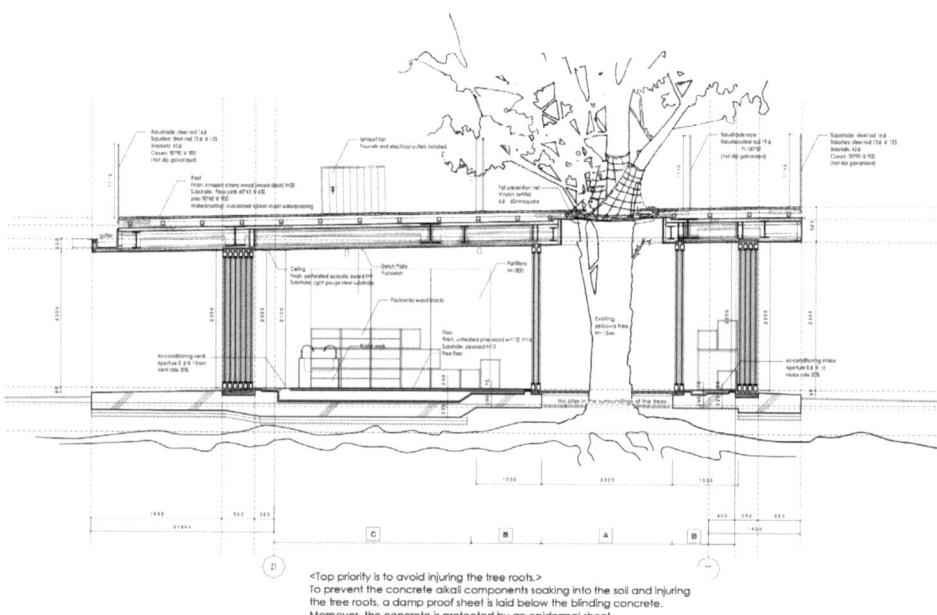

<Top priority is to avoid injuring the tree roots.>
To prevent the concrete alkali components soaking into the soil and injuring the tree roots, a damp proof sheet is laid below the blinding concrete. Moreover, the concrete is protected by an epidermal sheet.

Internal acoustic and thermal strategies are intentionally relaxed to incorporate background noise as a vital part of the experience, and promote a more dynamic thermal relationship with the changing weather beyond. This is framed further by careful incorporation of (and proximity to) existing Zelkova trees, in addition to the form of the building—which encourages interactive play and constant active movement. The consistent internal and external facades are fully openable to allow natural cross ventilation, while solar shading is provided by generous eaves. Below, Korean style under floor heating supplies the continuous interior through warm air convection in the winter.

The roof platform is a carefully curved subtly changing hyperbolic surface which, from a distance, appears to be flat. This is supported by a triangulated steel sub-structure which carefully negotiates its way around the existing trees, whilst echoing the buildings footprint, an irregular ellipse without a singular geometric center.

The ground slab is protected from tree root growth by an epidermal sheet, and the roots protected from alkali within the concrete seeping into the soil by a damp proof course, thus ensuring the health and long term relationship between trees and students.

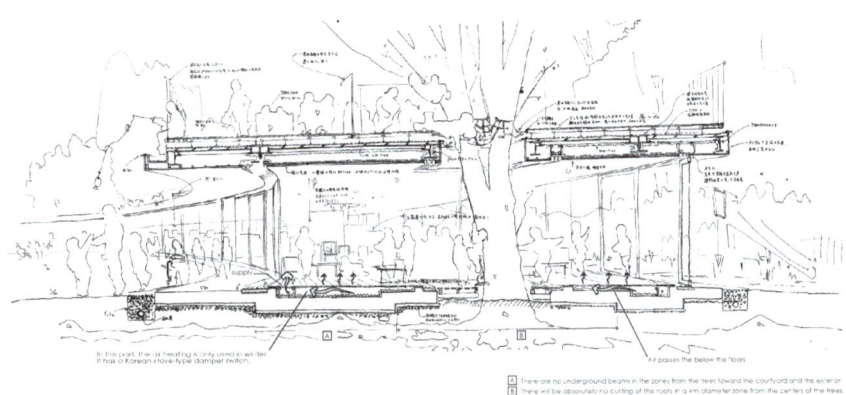

fig.3_10: [Left] Typical Construction Section

fig.3_11: [Above Top] Aerial view showing rooftop play area

fig.3_12: [Above] Sketch Section, addressing heating system

## Lycée français Alexandre Yersin, Hanoi

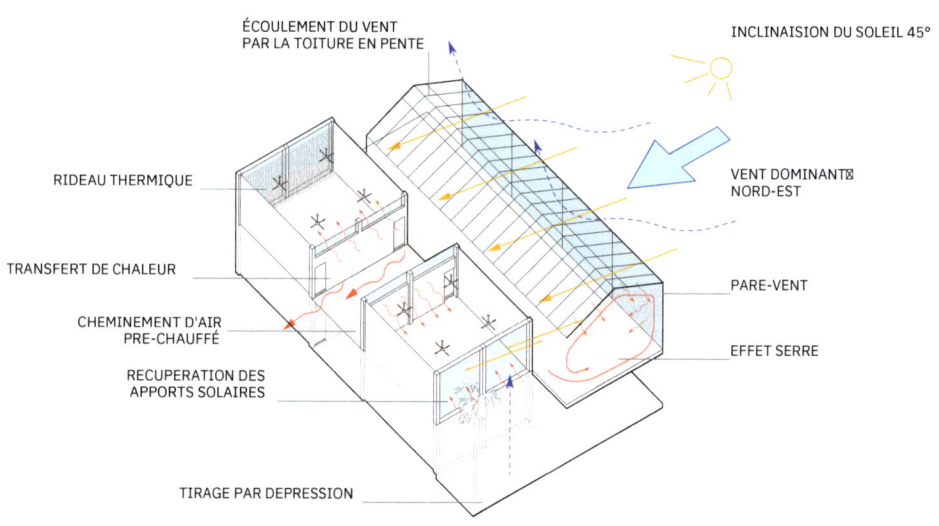

Situated in the subtropical climate of Hanoi Vietnam, this project is adapted to its environment. A lightweight structure of two densities differentiates between the very lightweight outer framework and the more supportive inner framework, allowing for a slightly more controlled intermediary microclimate in between. The two axonometric diagrams discuss an environmental strategy that allows for interactive use of screens and direct control of airflow, extensive use of planting within the perimeter of the building also helps to manage high levels of humidity. This system efficiently manages sun, wind, and rain parameters across the different seasons.

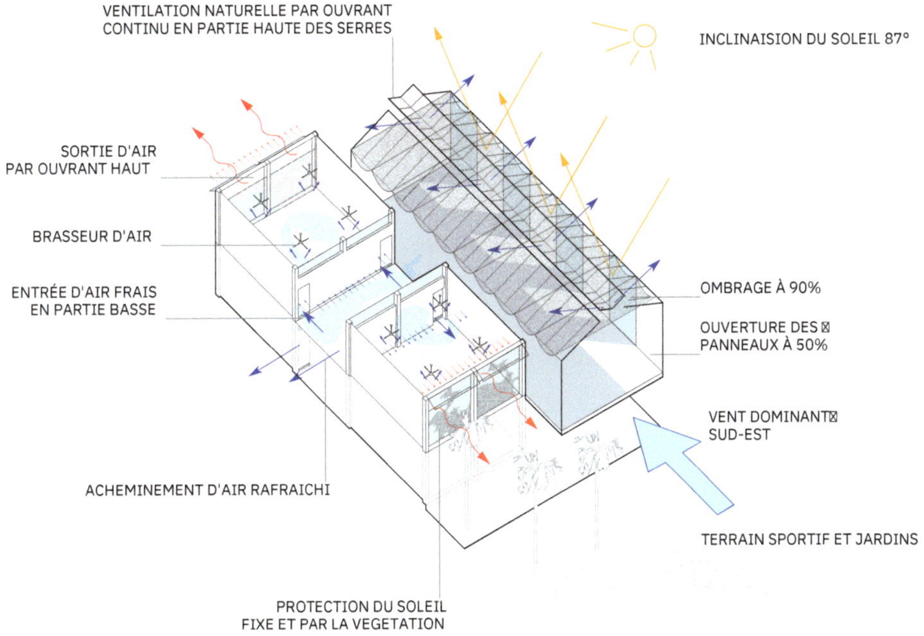

fig.3_13: [Left] Distributing interior warmth from solar gains in the winter

fig.3_14: [Left] Distributing interior warmth from solar gains in the winter

fig.3_15: [Above] making use of natural ventilation flow and efficient solar shading during the warmer seasons

## The EPFL Learning Center proposal, Lausanne

The EPFL Learning Center proposal (2004), by Valerio Olgiati, utilises a carefully aligned arrangement of 'A' shaped columns positioned to resolve a complex array of lateral forces created by the leaning columns to the perimeter. The 'A' columns prevent shear movement and thus avoid large torsion (twisting) forces at the center '+' sectioned columns.

*'The great hall on the ground floor of the new Learning Center is entered from the ramp. This will create a new central hub for the campus, where people will meet to work and talk, to sit and relax, or merely while they are on their way to somewhere else. All the various activities that enliven the campus will intersect here. An escalator will connect directly with the libraries; a ramp will lead to the Lecture Theater, while a spiral staircase will lead to the restaurants, the Learning Center and the bookshop. The building's frame, which features considerable spans between its supports, is made from in-situ concrete, colored reddish-brown. It is designed for maximum flexibility, with the very wide spans allowing for large internal spaces. The powerful concrete structure makes its presence felt on every floor and functions like a foundation for the entire building.*

*The structure itself—irregular and seemingly arbitrary in its constellation of forms—is the outcome of precise structural calculations. The result is a dynamic yet coherent three-dimensional figure. Each section, even when it is only partly visible, is clearly an element of the whole. The building thus forms a single, cogent entity.'*

Excerpt of original description by Valerio Olgiati. © Archive Olgiati.

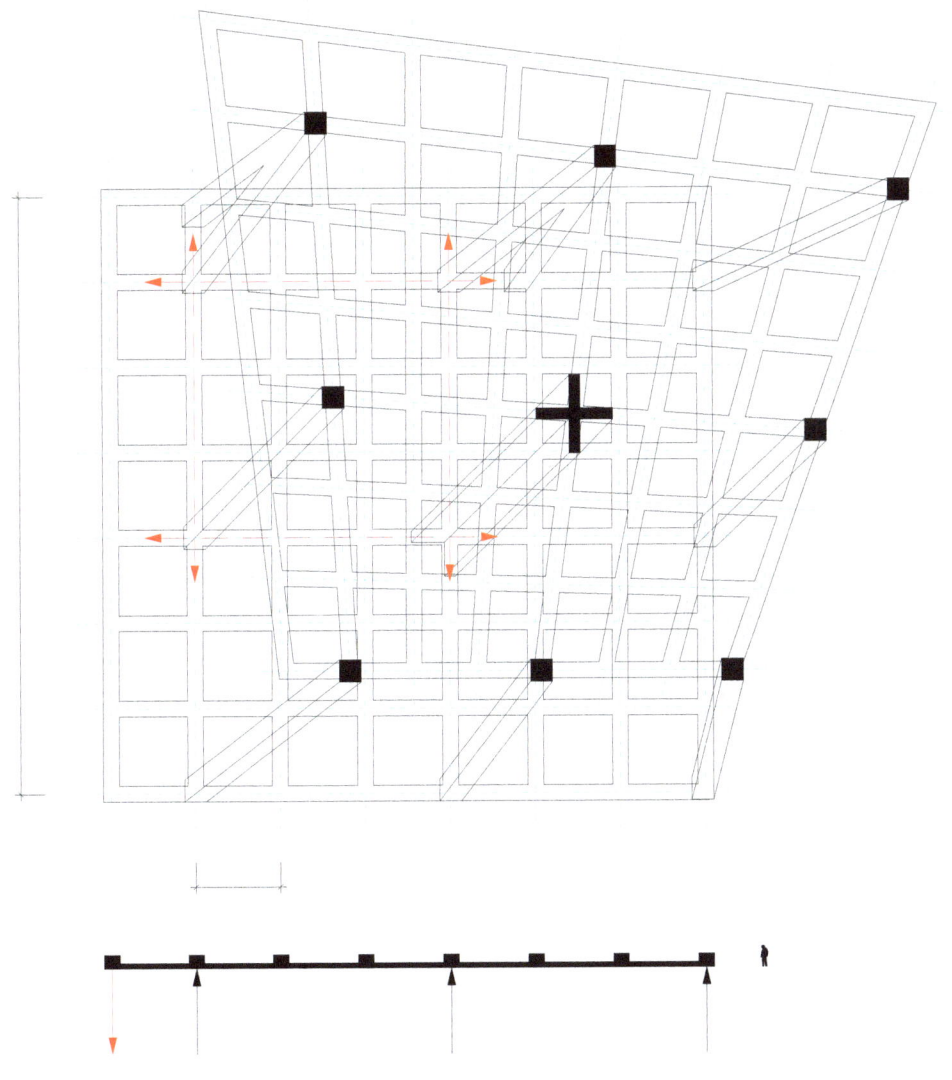

fig.3_16: [Left] EPFL Learning Center proposal, Structural model
fig.3_17: [Above] EPFL Learning Center proposal, Structural strategy diagram

The Table, Tokyo

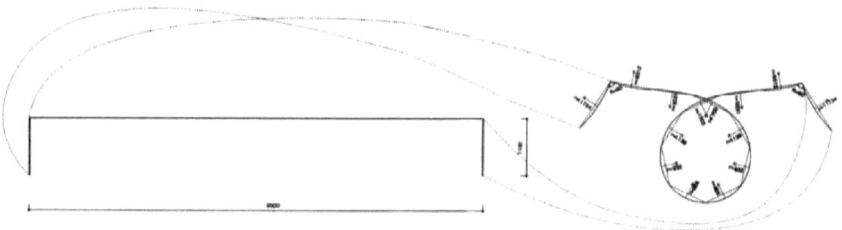

Designed by Junya Ishigami. A playful interaction between structural engineering and furniture design, the Table is in reality both an art work and a thought. Here Ishigami makes good use of 'pre-cambering' to deform a 9.5m x 2.6m x 3mm sheet of steel, before reshaping it back to a flat plane and maintaining this state through an array of carefully positioned, and carefully weighted, table-top place-setting objects. It is the distribution of objects in opposition to the materials new inclination (to return to a curved state) that creates a balanced equilibrium.

As it needs only the traditional four legs (also pre-cambered) to stand Ishigami here calls on the materials technical potential to create a seemingly impossible magic realist transformation of the familiar notion of 'table'.

*fig.3_18: [Left] The idea of the Table*

*fig.3_19: [Above] Pre-cambering the Table*

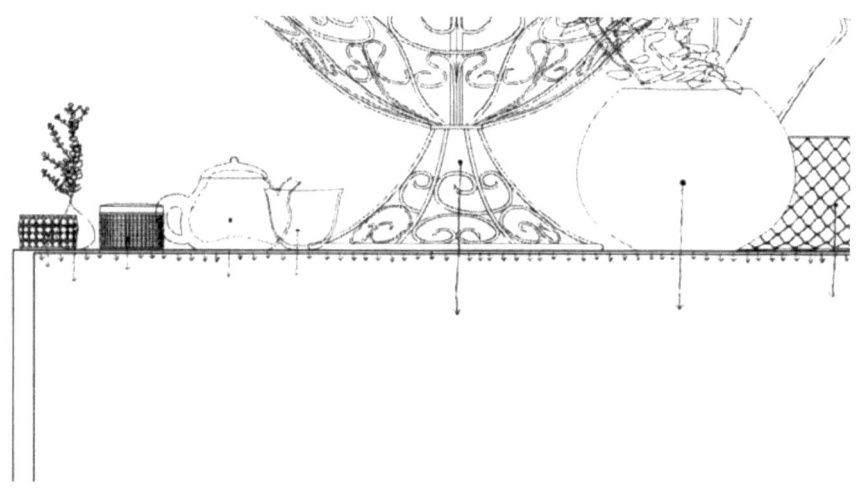

fig.3_20: Precision loading the table to create equilibrium
fig.3_21: [Right] The Table installed

*4. Matter matters*

## 4. *Matter matters*

> *'This is the craftsman's proper conscious domain; all his or her efforts to do good quality work depend on curiosity about the material at hand.'*- Richard Sennett [1]

For all the talk about 'space', matter matters—in the sense of materials. Moving from the ancient to the contemporary what follows is a series of points, focussing on material itself. Wood, stone, rammed earth, concrete, brick, ceramic, glass, metals, bamboo, and plastics. Followed by the 'new' interpretations of many of these same materials. The aim is to encourage a small shift in our understanding of material qualities; qualities occasionally taken for granted.

Commencing the list, some of these points may already be familiar:

**Wood** is thought to have emerged 400 million years ago.[2] There are now approximately 60 065 different species of trees[3], all of which can be classified as either Softwood or a Hardwood *(Fig.4_2–6)*. Hardwoods are typically stronger than Softwoods. However, Cork and the soft Balsa we use for model making are both Hardwoods.[4] The term hardwood relates to the cell structure and seed type of the timber. A further distinguishing factor is that Softwood trees are usually Evergreen (Conifers), whereas the other softwoods or hardwoods sheds leaves in the winter i.e are deciduous. Trees acquire over 90% of their nutrition from the atmosphere and the remainder from soil.[5] One Plane tree (common on London's streets) will absorb a tonne of $CO_2$ by the time it reaches 40 years of age.[6] This is also usually the minimum age of timber used in construction.[7] Trees affect rain.[8] A single mature oak tree can emit 400 litres of water a day into the atmosphere.[9] Strategic positioning of trees in relation to a building's facade can reduce air conditioning by up to 30%.[10]

The stiffness of timber is proportional to the amount of lignin, a class of complex organic polymer present in the support tissue.[11] Softwoods have a lower percentage of lingin and are usually lighter in colour. Softwoods account for about 80% of the world's timber production.[12] Pine is denser than some hardwoods but as it grows faster than many other timbers it makes an affordable alternative.[13] Typically, it takes 30 to 35 years for a hardwood tree to fully grow before harvesting, Oak typically needs 80 years. In Wales, over a period of six years from 1795, one person, Colonel Thomas Johnes, was responsible for planting over 900 000 Oak trees on his estate as part of a drive to bolster the country's ship building supplies.[14]

Perhaps contrary to common understanding some timbers, including Ironwood the hardest of all timbers, are heavier than water.

**Stone** is one of the strongest building materials available. The relationship between the words rock and stone is akin to the difference between tree and timber. The lightest rock is pumice, so light it can float on water—there is a pumice island or raft floating on the ocean near Tonga.[15] Basalt can be found on the Moon, Venus, and Mars.[16] It also happens to be the most common rock on the earth's surface and makes up large parts of the oceans floor.[17] Marble is found amongst the oldest parts of the earth's crust and starts off as limestone. Pulverised marble is used in pharmaceuticals (including Alka-seltzer) to neutralise stomach acid.[18]

In nature, Wurtzite Boron Nitride and mineral Lonsdaleite are technically the strongest substances at 17% and 58% respectively stronger than diamond.[19] However, diamond still holds the crown as by comparison these minerals are considered too rare. Clay, the basis of most ceramics, can be described as made of tiny particles, which come from the weathering of rocks and minerals and may also contain organic matter.[20] Like clay, sand is also a result of the weathering of rocks and on the particle size scale it sits between silt and gravel. Different material uses require different types of sand. For example, the desert sand of Dubai is unsuitable for water filtration, glass or any type of concrete as the grains are too smooth and so other appropriate sands are imported from as far as Australia for use in construction. As such sand is the world's most mined material, greatly exceeding natural renewal rates, with increasing environmental consequences.[21]

**Rammed Earth** consists primarily of soil, though sometimes with an added binder, that is manually compressed layer by layer between shuttering to form vertical walls. This is more of a method than a material, but one present since ancient times—so ancient it's how geological formations are made and why London is so good for tunneling.[22] It has the added benefits of being highly resistant to fire and sound, whilst being load bearing.[23] However, due to the nature of its construction, the minimum effective wall thickness is 300mm, which can have a sizeable impact on the floor area of small plan configurations. As a rule of thumb the maximum height for rammed earth construction is said to be defined as up to ten times the wall thickness.[24]

Stabilised rammed earth is one of the most environmentally-friendly and energy-efficient construction materials in the world. Typically, heat takes between 8 to 9 hours to move through from one side of a 300mm thick rammed earth wall to the other side. To make the most of this time delay solar passive buildings are typically designed to ensure that walls facing the sun do not get direct summer sunlight, whilst also ensuring that maximum direct winter sunlight is achieved for as long as possible.[25]

**Concrete**, the most widely used man-made material on the planet, is a mixture *(Fig.4_7)* of 60–65% aggregates like sand, gravel, and crushed stone; 15–20% water; and 10–15% cement. Select additives can be used to change setting time, decrease steel reinforcement bar (re-bar) corrosion, decrease viscosity, increase strength etc. Mix ratios are tuned for use. The concrete seen on a gallery floor, kitchen worktop, roof, foundation system, or bridge structure are not the same mix. Given the combination of water resistance, durability, strength and cost, concrete is now also one of the only materials we use for foundations, basements and tunnels; so we rely heavily on it. It is also commonly used for the hulls of floating homes. Once poured, it normally takes four weeks to reach full strength. Typically, for every 1 tonne of concrete made, 0.9 of a tonne of carbon-dioxide ($CO_2$) is produced.[26] Apparently, some proportion of this may be reabsorbed over its primary lifespan[27], though for typical modern reinforced concrete constructions this lifespan is typically thought to be only 50 to 100 years.[28]

**Brick** is not a raw material. Often taken for granted but not to be underestimated, brick requires an alternative approach to the idea of precision—there is an almost deceptive level of accuracy in bricklaying. For walls of up to 5m long and 5m high the maximum deviation (tolerance) is ±8mm, ±12mm for walls over 5m in length.[29] There are at least 13 named configurations of brick bond (or layout pattern), with more invented all the time *(Fig.4_8)*. Brick facade opening positions will depend on the type of bond chosen—as such it is common to set out brick spacing on plan section and elevation at the earliest stages of design. The standard UK brick dimension is 102mm wide, 65mm high, 215mm long though there are many alternative named brick sizes. Mortar thickness is typically specified at 10mm but can vary depending on design intention and bricklayers advice. The profile of a mortar joint should also be specified.

Bricks are made primarily from clay and are fired at 1 000 °C.[30] As such they can withstand extreme temperatures and are believed to be one of the most durable of all building materials although, if a wall is incorrectly designed, it is easy for water to explode brick fascias through freezing-thawing cycles in a process called spalding. The inverted triangular indentation atop some forms of brick is called a 'frog'. There is some debate about its purpose and whether it's better served laid up or down - though frog-up avoids air pockets over the mortar. Built in 1893 and standing at 60m the Monadnock building in Chicago is considered (at of time of writing) the tallest load-bearing brick building in the world.[31] Today brick is commonly used as an outer leaf wall layer with insulation and blockwork, timber, or other steel frameworks to form cavity walls. Construction clay bricks have a compressive strength that ranges from 5 to 125 Newtons/mm$^2$ depending on the type, age and manufacturing technique; London stock brick being the weakest at 5 to 25 Newtons/mm$^2$ and solid wire-cut bricks at 90 Newtons/mm$^2$, with class A engineering bricks at 125 Newtons/mm$^2$.[32] Elsewhere, it is believed there are presently enough Lego bricks in circulation for every person on earth to have 62 each.[33]

Looking around at the brick buildings in our environment, and the variety of bonds they use, it is occasionally worth reflecting that each and every brick was laid by someone by hand. As such, labour makes up a large proportion of the cost of building with brick.

**Ceramic**, described as *"a non-metallic solid material comprising an inorganic compound of metal, nonmetal or metalloid atoms primarily held in ionic and covalent bonds"*[34], is considered by some to be one of the earliest human inventions, with ceramic figurines found in the Czech Republic dating back to 29,000 BC[35], and ceramic vessels found in Jiangxi, China dating back to 18 000 BC.[36] Porcelain was first invented in China and in its unglazed form is as hard (Mohs scale ~7) as most granites (Mohs scale ~6.5). Ceramic tile has a high strength and is closely related to brick but it is more brittle. To stay clean in the thick smog or air pollution of early industrial London, the Natural History Museum was clad in ceramic tiles.

**Glass** is 100% recyclable and can theoretically be recycled endlessly without loss in quality or purity.[37] Recycling a glass bottle saves enough energy to light a 100 Watt light bulb for four hours.[38] Most glass bottles and jars contain at least 25% recycled glass.

Glass is made in nature when lighting strikes sand.[39] It is also known as the 'fourth state of matter' since, arguably, it is understood to always be in a supercooled liquid form—the molecules are just moving incredibly slowly—and as such it has no solid or gas state.[40] This is a subject of interest in Condensed Matter Physics[41] which, is also exploring the notion of making glass from metal.[42]

Glass as a window material was first used by the Romans mid 1$^{st}$ century AD,[43] but didn't reach UK housing until the 17$^{th}$ Century. Depending on the thickness of the glass, it typically weighs between 7.5kg/m$^2$ (3mm thick) to 50 kg/m$^2$ (20mm thick). Glass today is made into panes, or cast into structural blocks, it is otherwise brittle unless laminated (often with plastics) for extra strength. It is now possible to specify window glazing[44] with insulation properties similar to a typical brick cavity wall whilst simultaneously heating interior environments through transparent electrically conductive coatings such as Indium Tin Oxide (ITO).[45]

Fiberglass is made from the same substances as normal glass. It's heated and pulled into threads which are then woven together.

**Metals** make up 90 of the periodic tables 118 elements of which only 7, excluding their alloys, are used in specific instances in construction: lead; tin; zinc; copper; nickel; iron; and aluminium. Aluminium is thought to be the most prevalent metal in the earth's crust and Iron is thought to be the most prevalent within the earth itself. Iron mixed with carbon together make steel, of which approximately 1 600 million tonnes are produced annually.[46] Stainless steel additionally includes nickel and chromium—another metal, though it is very durable, it is also difficult to connect. Like the inherent protective (rust) coating of Cor-ten (mild Steel) zinc also reacts with the atmosphere to form a degraded zinc carbonate on its surface— which in turn protects the zinc beneath.

Not all are one metal: Bronze is a metal alloy made from copper and tin. Copper makes up the larger component in bronze, usually between 80 to 95%. Brass too is an alloy of copper and zinc. Copper and many copper alloys are thought to be highly resistant to bacteria—including current strains of antibiotic resistant superbugs—through the release of copper ions.[47] Silver is also believed to be antimicrobial. However, unlike copper silver requires water for its antimicrobial properties to take effect. Ionic exchange between different metals can also leads to galvanic corrosion.[48] Galvanic corrosion is caused when the wrong combination of metals are placed together in direct contact, and moisture of some form is able to interact between them. As such, there are tables that highlight which combinations of metals to avoid. This is important for example when detailing metal sheet cladding and its substructure fixings. A famous example of galvanic corrosion failure is the Statue of Liberty in which a reaction between its copper envelope and its iron substructure caused the iron to corrode and the fixings to subsequently buckle.[49] St Mary's Cathedral in Tokyo is another example of galvanic corrosion, where the stainless steel roof cladding peeled off in a storm after reacting badly with its carbon steel substructure.[50]

Most metals are formed of crystalline structures. This means there are inherent structural weaknesses (or creep) at the boundaries between crystals. Each of the small but visible shapes common in the texture of galvanised zinc is a single metal crystal *(Fig.4_9)*. For most uses the molecular bond between crystals is such that these boundary conditions are negligible. However, for some specific and extreme requirements new processes have been developed to cast large single crystals into whole components, with greater inherent stability and temperature resistance. An example of this are single crystal jet-engine turbine blades which can withstand temperatures of 2 000 °C without deforming.[51]

Wood pencils have never actually contained lead. Even though lead is soft enough it could be used for writing. Pencil lead is a type of graphite, and it was this same graphite —along with a piece of scotch tape—that helped to make the earliest forms of Graphene.[52]

The **Bamboo** *(Fig.4_10)* family is a part of the Grass family (not the Tree family) and has approximately 1 575 species [53] of which about 200 are thought to be used in construction.[54] Some can grow up to 0.8 meters a day (Moso and Madake Bamboo).[55] Other types of bamboo can grow for up to 120 years. It is thought that in the right conditions bamboo buildings can last for over 100 years. However, it is difficult to find effective case studies to verify this, as such 10–15 years with maintenance might be more realistic.[56] As a testimony to its resilience it was the only plant to have survived the Hiroshima bomb of 1945.[57] The tallest of species— Dendrocalamus giganteus aka Dragon Bamboo—reaches over 30m in height. Amongst others, the species of the Guadua genus are suitable for construction and typically need 3 to 6 years to grow fully before it can be harvested.[58] The Guardua genus has greater tensile strength than steel and in some situations can compete with concrete's ability to withstand compression; 40–80N/mm² for Bamboo compared to 25–80N/mm² typically for concrete—though pre-cast implementations of over 450+N/mm² have been developed for the latter.[59] However, its resistance to compression can decrease over time.[60] In 1880, Through sheer determination Thomas Edison tested some 1 600 material options and combinations in his search of the best incandescent filament. He eventually arrived at a carbonised form of bamboo.[61] Bamboo shoots are edible and bamboo root systems are good for stabilising and preventing soil erosion. Its carbon sequestering abilities, though much lauded, have been called into question as it may appear it stores less carbon than its equivalent ground area in trees.[62] It has also been suggested that bamboo could be a emitter of $CO_2$.[63]

**Plastics** began to make an impact in the West in the 1950s. Since then, it is believed 8.3 billion tonnes of plastic has been made.[64] That figure is similar in weight to 24 000 Empire State Buildings. Two-thirds of the 8.3 billion tonnes of plastic that has been produced since it was invented is thought to have been disposed of in landfill. In 2016 alone, 335 million tonnes of plastic was produced.[65] Approximately 5–12 million tonnes of that 335 million tonnes is disposed of in the oceans yearly.[66] The Pacific Ocean is host to a largely floating collection of plastic refuse believed to be larger than the size of France or about 3 times the area of the UK.[67] Plastic can remain in the environment for many hundreds of years, intact and mummified in landfill, or broken down into particles by sun and sea through which it can enter the food chain.[68] Recent research has found that mealworms and their gut bacteria are happy to eat polystyrene, creating a bi-product that's safe to use as crop soil.[69] The widespread use of plastics and polyurethane foam in modern furnishings is a significant contributing factor to house fires spreading 6 times faster than they did 50 years ago.[70] Poly Vinyl Chloride (PVC), a common form of vinyl was discovered accidentally.[71] It was discovered twice, in 1835 and then again in 1872 but not perfected until 1913.[72] PVC has an effective lifespan of up to 100 years.[73] Due to its compressive strength and insulating properties Expanded Poly-Styrene (EPS) makes a good accompaniment to concrete, whether as sacrificial formwork, or as warm roof (also known as inverted roof) insulation.[74] Plastics can be used to improve visibility, or to glow in the dark. They are lightweight. They can integrate with other materials (such as laminates in glass), form valuable building components, synthesise complex forms and withstand a variety of extreme environmental conditions. New plastics research is seeking ways to reduce our reliance on other forms of energy expenditure such as highly reflective porous polymer coatings able to deflect over 90% of sunlight, and potentially applied in the form of paint.[75] This quality of plastics when used in construction suggests the possibility of avoiding mechanical ventilation if applied to facades in the right scenarios—it would be necessary to consider the impact of reflected heat and light on the surrounding environment. About now might be a good moment for a tea break.

**New woods** have enabled new properties to emerge, through innovative approaches to processing and manufacturing, in this otherwise old material. Beyond the mid-20th century developments of plywood and chipboard, for example, it is now possible to make wood transparent, so as to increase its tensile strength to make it stronger than steel.[76] Techniques for augmenting new woods are capable of forming insulation that's more efficient than many forms of polystyrene based insulations.[77] In another approach, transparency is made possible by removing the lignin and replacing it with epoxy resin. Super strong wood is achieved by boiling it in a chemical solution similar to part of the process of making paper. This also partially reduces its strength. Then the wood is compressed until the cell structure collapses whilst being heated until new chemical bonds are created. This manufacturing process also triples the density of the new woods.[78] Insulating 'Nanowood's properties are thermally unidirectional and biodegradable. These processes are energy intensive and not presently scaled for mass production.

In addition, longer established forms of 'engineered timber' including Cross Laminated Timber (CLT), Pre-stressed Timber and Glulam—a similar construction to CLT, although the grain of its constituent parts are aligned in the same direction

– have built on the legacy of previous mid-century innovations in Chipboard (Particle Board) and Plywood manufacture. These developments in wood have and are continuing to take construction processes in new directions. Twelve year old trees specially farmed in sustainable carefully managed forests are suitable for these forms of timber. In addition to the use of pre-stressed timber for increased earthquake resistance in buildings, Glulam and CLT allow monolithic construction, taller buildings, and wider spans to be achieved.[79]

**New concretes** are, like new woods are to timber, enhanced performance versions of concrete. Some forms of Roman concrete, invented over 2 000 years ago, have been discovered to possess 'self-healing' properties transmitted through the combination of lime and volcanic ash.[80] This ash contained a rare mineral known as aluminium tobermorite. Instead of causing erosion, exposure to seawater causes aluminium tobermorite to crystallise in the lime as it cures.[81] Ancient lessons like these are being reconsidered to bolster the properties of new concrete. Alternative contemporary interpretations of 'self-healing' adapts limestone producing bacteria and fungi to repair small cracks (micro-fissures) as they occur.[82] Other approaches towards greener concretes use forms of ash such as Fly-ash, a common by-product of power stations, also known as 'pulverised fuel ash' or PFA.[83] PFA also creates a stronger concrete whilst reducing $CO_2$ emissions by up to 70%.[84] But this is not without its drawbacks which include an increased salt efflorescence (potentially problematic for steel reinforcing corrosion), concerns about cold climate performance (freezing–thawing cycles), and the need for more time to gain full strength.

Further developments in new concrete include Eco-friendly Ductile Cementitious Composite (EDCC) which can flex in a manner similar to steel.[85] EDCC is therefore suitable for earthquake prone regions and is already in use in areas at risk from earthquakes in Canada and India. New lightweight concretes have been, and are being, developed to build tall structures. Elsewhere, polymer-concretes combine some of the advantages of both materials. There is also a concrete designed to clean air by absorbing nitric oxide (NO) and nitrogen dioxide ($NO_2$), two compounds common in urban air pollution.[86]

**New ceramics** have augmented old ceramics and can now be found as specialist variations in mobile phone bodies, high performance disk brakes, high voltage electrical insulators, electrical superconductors, high performance thermal insulation (space shuttle heat shields), precision cutting tools, and dental fillings.

**New composite** combinations of different materials are constantly being explored and invented. In most cases, it is important to consider what, and how, composite elements or aggregates are held together. Glue for example makes up between a quarter and a third of Carbon Fiber, an advanced material usually found in high performance mechanical systems and used to reduce the need for movement joints in poured cementitious flooring, and the making of some bathtubs. Newer composites include alternative mixtures of Graphene with concrete, ceramics, and other materials mentioned here.[87]

Further to all this there are new and old material treatments that can: avert water via oleophobic coatings; or use water to expand in volume (known as hygroscopic swelling); use temperature to change shape as in bimetallic strips; use temperature to change colour through thermochromic inks; use sunlight to change colour with photochromic compounds; use sunlight to generate energy through photovoltaic and perovskites cells; convert energy to light via electroluminescent paint; use phosphorescent compounds to store sunlight and later illuminate in the dark; and piezoelectric materials that convert movement to energy.

**Refuse** and the context of our environment makes it important to consider the byproducts of processing, shaping and manipulating materials. The UK construction industry alone uses 420 million tonnes of material a year, of which a quarter becomes waste. This generates almost a third (32%) of total landfill waste and almost a fifth (19%) of the UK's greenhouse gas emissions.[88]

**Process** is key. Performance properties for any given material can vary dramatically depending how it's cut, oriented, mixed, fired and so forth.

**Elements** encompass this world of matter with characteristics of their own:
*Water*—the innocent villain: can take advantage of temperature and air pressure to cycle through three of its four*[89] states to; freeze into ice crystals within brick or stone causing the material to split (spalling), or through condensation or infiltration enable bacteria to cause mould on surfaces or rot in wood. Or sequentially infiltrate cracks and condense on reinforcement bars to encourage rust and induce spalling in concrete—sometimes seemingly deliberately.

*Fire*: if given the opportunity, can make use of the optimum air conditions in ducts, wall or facade cavities to spread across rooms and floors unless properly compartmentalised. In most cases the material by-products of smoke and reduced oxygen are more dangerous than the flame itself.

* Discovered in 2016 and known as 'Tunneling' water molecules exhibit quantum behaviour in this fourth state when pressurised and confined in extremely tight spaces. This state is found in nature in the gems aquamarine and emerald.

*Air*: typically 21% oxygen; 78% nitrogen; <1% $CO_2$, can partner with water moisture to carry mould spores as well as other pollutants. As wind it applies lateral forces to structures, but is just as happy to apply lifting forces to pull roofs up.

*Sunlight*: or more specifically its ultraviolet component, can weaken material surfaces at a molecular level, eventually fading colours or weakening unprotected (typically synthetic) membranes.

These are disruptive effects of typically life sustaining elements. If overlooked they can work in tandem to undo material arrangements large or small. Or, if looked into, a deeper understanding will support critical decisions when arranging materials.

### Conclusion

*'...Given that the desire to develop new ideas is reinforced by new technical opportunities, by new formal liberties and a common thirst for novelty and identity, we have a more profound responsibility to define what is appropriate and meaningful. Choice is no replacement for quality, abundance of ideas is no replacement for precision of thought, novelty is no guarantee of rigour.'* (Chipperfield, D. 2018) [90]

This stream of often haphazard, occasionally sporadic points, was a primer on a few of the more peripheral qualities of commonly used materials. This is also a small introduction to the greater body of publications and research papers providing useful insights into the rich world of material sciences. Do not be tempted to adopt any of these points as the sole basis of a design project because to do so would limit the potential and depth of your TS. Instead, consider how the 'idea' of the material might coordinate with a design project's larger agenda, not to mention structural and environmental strategies.

Pragmatic guidance on the architectural application of these materials may be found in the texts referenced in chapter 2 'The Field'—especially the first two sections: 'Overview', and 'Construction'. In particular The Construction Materials Manual, and the Why Materials Matter books (also referenced) both provide a lead-in towards more specialist books. The former is itself part of a series with a good specialist book on each of the materials discussed here.

Studying both traditional and contemporary knowledge of materials will go a long way towards understanding and communicating the physical scope of a project. Experimentation will cover the remaining distance and instil a more tacit understanding of select materials. Beyond this text, and beyond the measurable, there are further perspectives that find pleasure in exploring the ontological social and cultural meanings of materials. Considering the innate qualities of materials in these ways may help us better appreciate the rich palette of matter from which we compose our built environments.

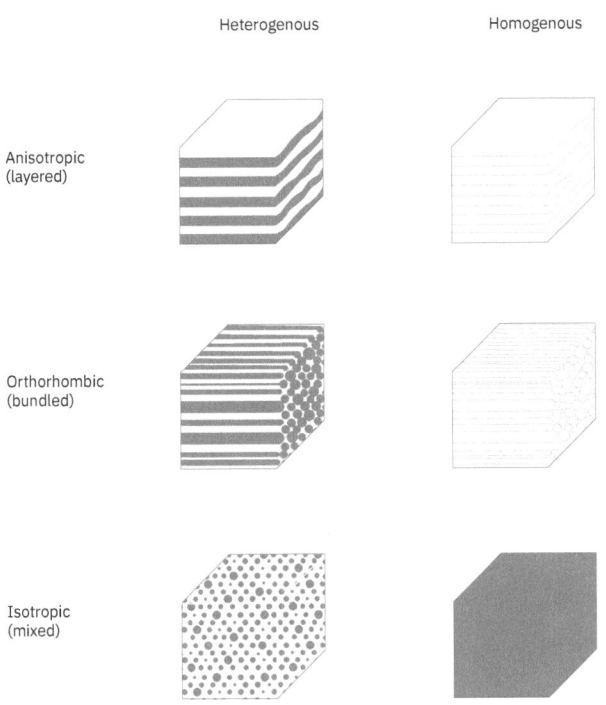

Heterogenous materials are made of more than one element, whereas homogenous consists of the same element throughout. In both anisotropic and orthorhombic materials one axis has different properties to the other two, orthorhombic can appear as anisotropic if viewed from the same perspective. Isotropic materials exhibit the same properties across all axis. These terms are typically used to describe the micro composition of a material. In many cases the most appropriate description might change if the scale shifts, i.e. at a larger scale of view a heterogenous material may appear homogenous.

*fig.4_1: Material composition descriptors*

## Woods close-up

The walls of tracheid (fibre) cells (a. and b.) consist mainly of lignin—providing firmness, and cellulose—providing toughness. The core of these tubular fibres also contains small amounts of protoplasm (living tissue) but are otherwise used as conduits for circulating liquid sap (water and nutrients) throughout the trunk. Tracheids make up approximately 90% of softwood, with paranchyma (d.)—food storage—making up the bulk of the remainder. Resin is typically only found in softwoods and resin ducts (c.) are usually too small to see by eye. Hardwood is categorised as either ring-porous, or diffuse-porous, depending on the vessel sizing and distribution.[91]

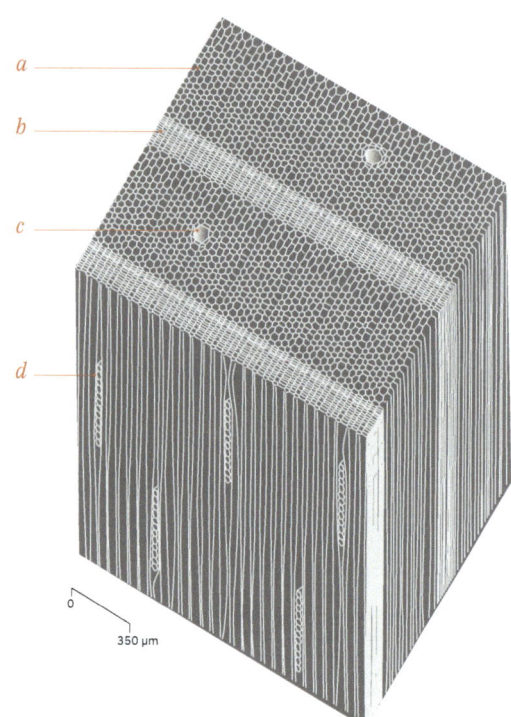

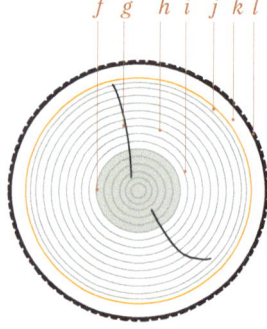

*Trunk section*

Hardwood and Softwood elements:
- *a.* Earlywood tracheid fibres (springwood)
- *b.* Latewood tracheid fibres (summerwood)
- *c.* Resin duct. Softwoods only
- *d.* Ray cells. Consisting of small paranchyma cells aligned transversely horizontally towards the centre of the trunk.
- *e.* Vessel (proportions exaggerated). Hardwoods only

Trunk cross section (Hardwood/Softwood):
- *f.* Heartwood is the core of a tree. It has ceased to carry sap, is typically denser, and forms a stronger wood than that surrounding it. In hardwood trees it usually contains tannin's and other organic chemical compounds making it more resistant to decay, in softwoods it usually contains resins.
- *g.* Knots, present in all woods, are former branches which have been subsumed by the growth of the surrounding sapwood.

*fig.4_2: Hard/Softwood tree trunk section diagram NTS*

*fig.4_3: [above right] Softwood nanoscale diagram*

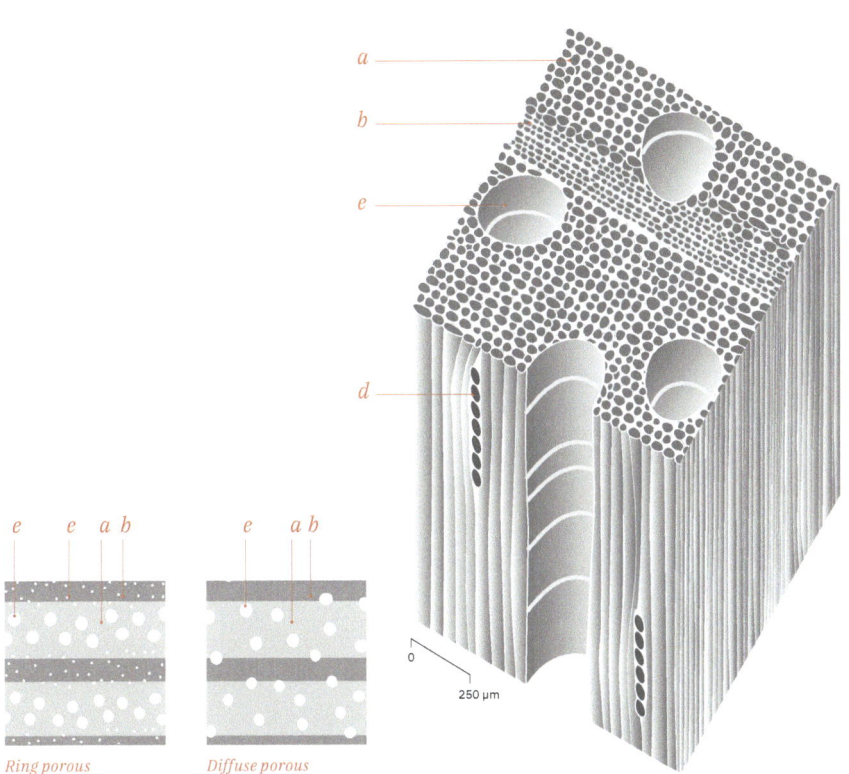

*Ring porous*

*Diffuse porous*

*h.* Annual rings reflect the change in growth rates of spring and summer sapwood as created by the cambium layer.

*i.* Sapwood constitutes the bulk of the surrounding timber contains the active tissue, sap (water and nutrient) transit, and secondary structure (latewood) rings.

*j.* Cambium is a very thin outer layer of highly active cells. It creates sapwood during the growth cycles and could be thought of as the wood production line.

*k.* Bast carries a sugar rich downward flow of sap from the branches and feeds the cambium which it surrounds.

*l.* Bark is a tree's outermost protective surface layer. It is generated and renewed by its own bark cambium under-layer and protects the body from insects, animals, weather and other risks. It also allows the core to breathe whilst retaining moisture.

*fig.4_4: [above left] Diffuse porous hardwood diagram NTS*

*fig.4_5: [above centre] Ring porous hardwood diagram NTS*

*fig.4_6: [right] Hardwood (diffuse porous) nanoscale diagram*

# Concrete

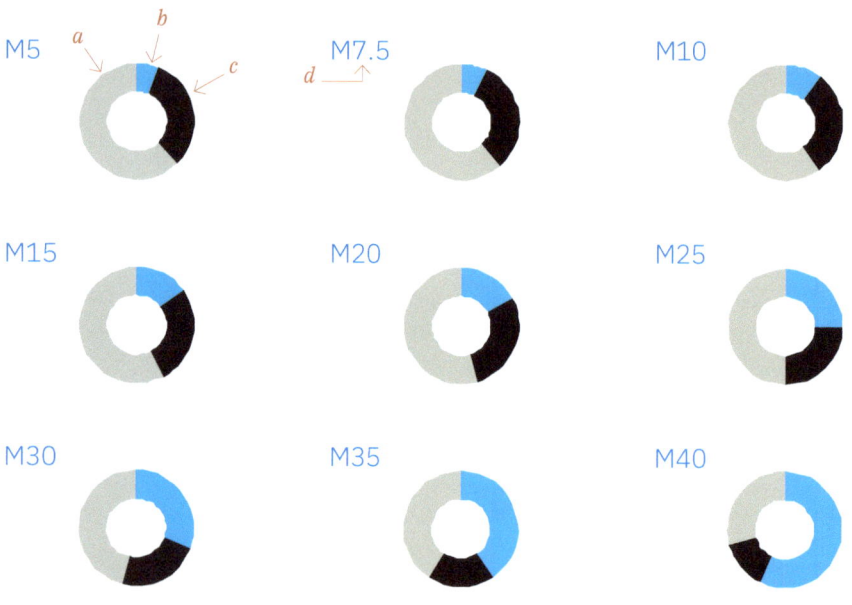

a. Aggregates
b. Cement (Binder)
c. Sand [the amount of sand used is half that of the Aggregate]
d. M7.5 = 'Mix' 7.5 N/mm² compressive strength > 28days

| Mix | Cement | Sand | Aggregates | Cement | Sand | Aggregates |
|---|---|---|---|---|---|---|
| M5 | 1 | 5 | 10 | 6% | 31% | 63% |
| M7.5 | 1 | 4 | 8 | 8% | 31% | 62% |
| M10 | 1 | 3 | 6 | 10% | 30% | 60% |
| M15 | 1 | 2 | 4 | 14% | 29% | 57% |
| M20 | 1 | 1.5 | 3 | 18% | 27% | 55% |
| M25 | 1 | 1 | 2 | 25% | 25% | 50% |
| M30 | 1 | 0.75 | 1.5 | 31% | 23% | 46% |
| M35 | 1 | 0.5 | 1 | 40% | 20% | 40% |
| M40 | 1 | 0.25 | 0.5 | 57% | 14% | 29% |

*fig.4_7: Concrete mix ratio's and compressive strengths*

## Brick

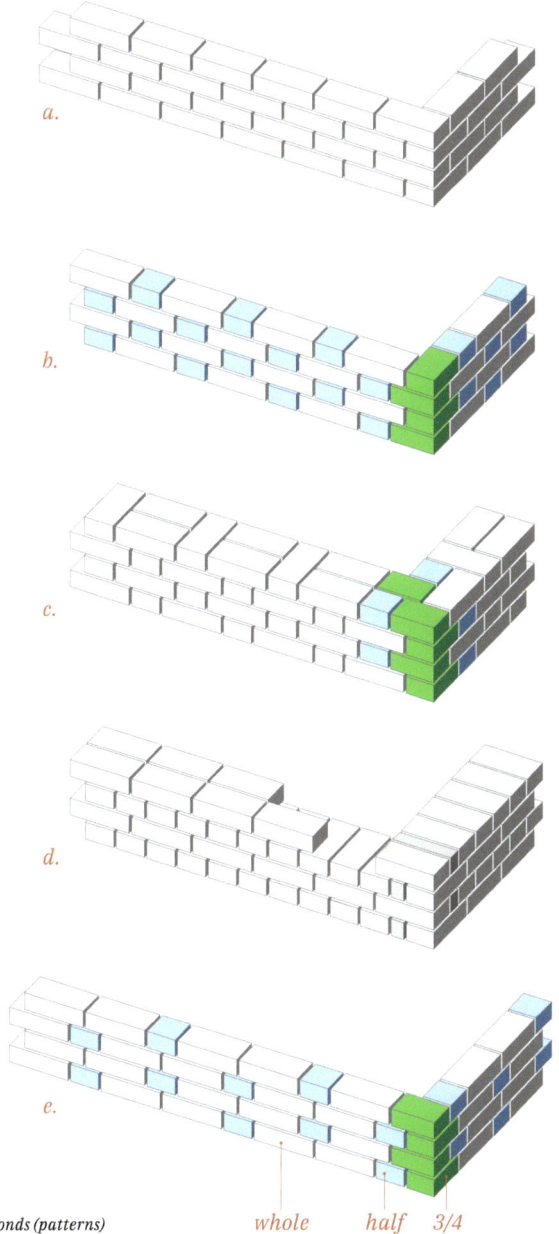

*a.* **Stretcher Bond**, the simplest form of staggered brick bond, all bricks remain whole but as this configuration is only one leaf (layers of brick) thick it is necessary to 'tie' (a number of small connections) the wall to an inner structural layer—typically blockwork. One of the most common bonds and the more affordable.

*b.* **Mock Flemish**, similar to Stretcher Bond but with a half brick (Blue) between every whole. More expensive than Stretcher due to additional labour of striking (cutting) the bricks.

*c.* **Flemish**, shares the same surface pattern as Mock Flemish but is two leaves thick and, with the exception of the corners, is made of whole bricks. Requires ¾ length bricks (Green) and half bricks at the corners. This configuration was typically used for load-bearing walls in older buildings before the use of Blockwork and Insulation. If retrofitting with insulation be considered about applying insulation to the inside as this can cause the outer surface to fail (spalding) in some climates.

*d.* **English Bond**, different courses (stacked lines) rotate to create an alternating pattern. As for Flemish this is an expensive bond.

*e.* **Monks Bond**, similar to Mock Flemish, can be made with a combination of half and whole bricks, or as two leaves for increased structural stability. Included here for its rarity.

*fig.4_8: Common and uncommon Brick bonds (patterns)*

*whole*   *half*   *3/4*

71

Metal

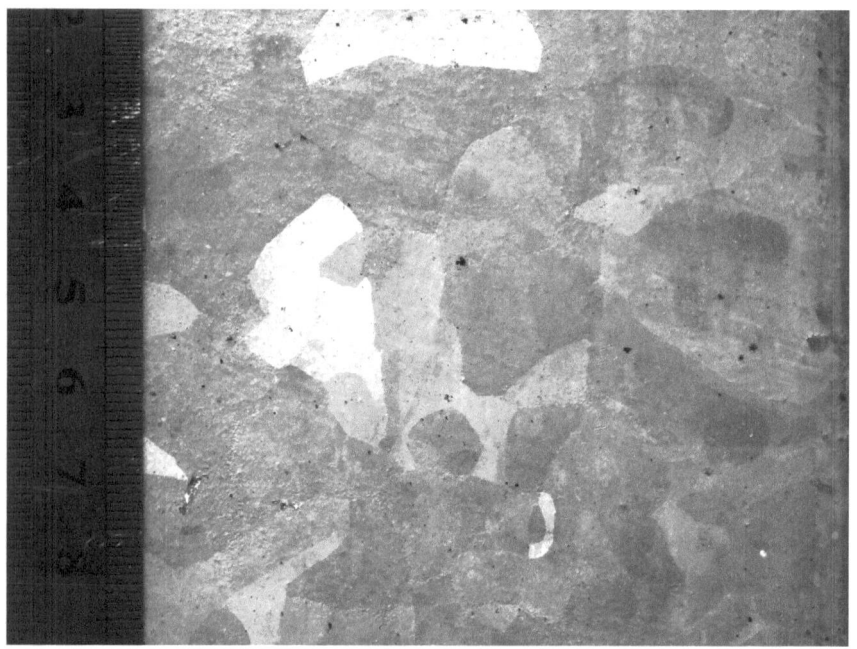

This example of a typical galvanised zinc coating reveals a visible form of amorphous crystallisation in metal. Each grey region here represents an individual Zinc crystal.grain, the thicknesses of typical zinc coatings range from 5 to 110 μm.

*fig.4_9: Visible galvanised Zinc crystals*

Bamboo close-up

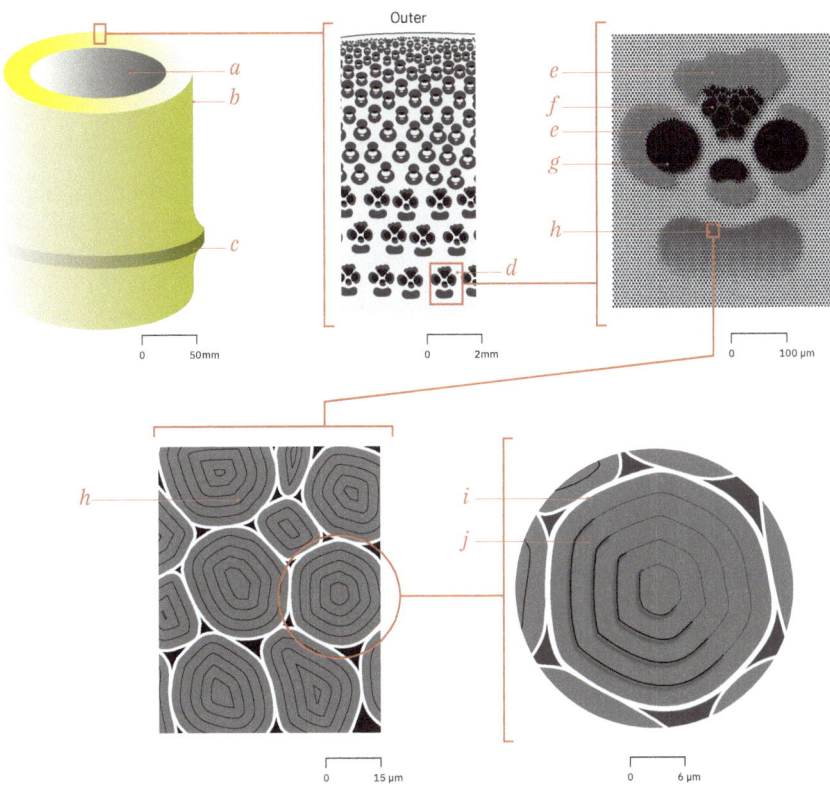

Bamboo is itself a feat of bio-engineering. Here we see the hierarchical arrangement of densely packed cellular fibres (lamellae) themselves composed largely of lignin and hemicellulose.

These combine to form larger bundles varying in density in relation to their proximity to the outer surface.

*a.* Lacuna
*b.* Internode
*c.* Node
*d.* Vascular bundle
*e.* Fibre sheath
*f.* Phloem and Sieve tubes
*g.* Metaxylem
*h.* Fibre bundle: Groups of 7 - 10 Lamellae fibres
*i.* Lamellae cell wall
*j.* Crystaline cellulose microfibrils

*fig.4_10: The nanoscale hierarchical structure of Bamboo*

# *Notes*

1. Sennett, R. (2008). *Craftsman, The*. New Haven: Yale University Press. ch.4, pp.120.
2. Christine Strullu-Derrien, Paul Kenrick, Paul Tafforeau, Hervé Cochard, Jean-Louis Bonnemain, Alain Le Hérissé, Hubert Lardeux, Eric Badel, *Botanical Journal of the Linnean Society*, Volume 175, Issue 3, 1 July 2014, pp. 423–437, https://doi.org/10.1111/boj.12175; "The earliest timber and its hydraulic properties documented in c. 407-million-year-old fossils using synchrotron microtomography."
3. E. Beech, M. Rivers, S. Oldfield & P. P. Smith (2017) GlobalTreeSearch: The first complete global database of tree species and country distributions, *Journal of Sustainable Forestry*, 36:5, pp.454-489, DOI: 10.1080/10549811.2017.1310049
4. Ilic, J., 1991. *CSIRO atlas of hardwoods*. Springer-Verlag.
5. Hershey, D. R. (1991). Digging Deeper into Helmont's Famous Willow Tree Experiment. *The American Biology Teacher*, 53:8, pp.458–460. https://doi.org/10.2307/4449369
6. McPherson, E.G. and Simpson, J.R., 1999. *Carbon dioxide reduction through urban forestry: guidelines for professional and volunteer tree planters*. Gen. Tech. Rep. PSW-GTR-171. Albany, CA: US Department of Agriculture, Forest Service, Pacific Southwest Research Station. 237 p., 171.
7. Ramage, M.H., Burridge, H., Busse-Wicher, M., Fereday, G., Reynolds, T., Shah, D.U., Wu, G., Yu, L., Fleming, P., Densley-Tingley, D. and Allwood, J., 2017. The wood from the trees: The use of timber in construction. *Renewable and Sustainable Energy Reviews*, 68, pp.333-359.
8. Sheil, D. and Murdiyarso, D., 2009. *How forests attract rain: an examination of a new hypothesis*. Bioscience, 59(4), pp.341-347.
9. Petersen, J.F., Sack, D.I. & Gabler, R.E., 2017. *Physical geography*, Boston, MA: Cengage Learning. P.149
10. Huang, Y.J., Akbari, H., Taha, H. and Rosenfeld, A.H., 1987. The potential of vegetation in reducing summer cooling loads in residential buildings. *Journal of climate and Applied Meteorology*, 26(9), pp.1103-1116
11. Young, R.A., 1985. The chemistry of solid wood. *Wood Science and Technology*, 19(1), Ch5, pp 211–255
12. Arets, E.J.M.M., Van der Meer, P.J., Verwer, C.C., Hengeveld, G.M., Tolkamp, G.W., Nabuurs, G.J. and Van Oorschot, M., 2011. *Global wood production: assessment of industrial round wood supply from forest management systems in different global regions* (No. 1808). Alterra Wageningen UR. pp 15.
13. Www5.csudh.edu. (2018). *PHYSICAL PROPERTIES OF COMMON WOODS*. [online] Available at: http://www5.csudh.edu/oliver/chemdata/woods.htm [Accessed 15 Dec. 2018].
14. Cohen, S. (2004). *Planting nature*. Berkeley: University of California Press.
15. Discovery News. (2018). Mystery Rock Shelf Floating in Pacific: *Discovery News*. [online] Available at: https://web.archive.org/web/20120815185904/http://news.discovery.com/earth/pacific-pumice-island-120813.html [Accessed 15 Dec. 2018].
16. Lawrence, S.J., Kramer, G.Y., Center, B.F., Jolliff, B.L., Hawke, B.R., Robinson, M.S., Hagerty, J.J., Taylor, G.J., Plescia, J., Garry, W.B. and Stopar, J.D., *Sampling The Age Extremes Of Lunar Volcanism: The Youngest And Oldest Lunar Basalts*.41st Lunar and Planetary Science Conference (2010)
17. King, D. (2018). *Basalt: What Is Basalt, How Does It Form, and How Is It Used?*. [online] geology.com. Available at: https://geology.com/rocks/basalt.shtml [Accessed 15 Dec. 2018].
18. King, D. (2018). *Marble: A non-foliated metamorphic rock that forms when limestone is subjected to heat and pressure*. [online] geology.com. Available at: https://geology.com/rocks/marble.shtml [Accessed 15 Dec. 2018].
19. Pan, Z., Sun, H., Zhang, Y. and Chen, C., 2009. Harder than diamond: superior indentation strength of wurtzite BN and lonsdaleite. *Physical review letters*, 102(5), p.055503.
20. Guggenheim, S. and Martin, R.T., 1995. Definition of clay and clay mineral: joint report of the AIPEA nomenclature and CMS nomenclature committees. *Clays and clay minerals*, 43(2), pp.255-256.
21. Peduzzi, P., 2014. Sand, rarer than one thinks. *Environmental Development*, 11, pp.208-218.
22. Jaquin, P.A., Augarde, C.E. and Gerrard, C.M., 2008. Chronological description of the spatial development of rammed earth techniques. *International Journal of Architectural Heritage*, 2(4), pp.377-400.

23. Houben, H., Guillaud, H. and Hall, B.B., 1994. *Earth construction: a comprehensive guide*. London: Intermediate Technology Publications, p. 362.
24. Bui, Q.B., Morel, J.C., Hans, S. and Meunier, N., 2009. Compression behaviour of non-industrial materials in civil engineering by three scale experiments: the case of rammed earth. *Materials and structures*, 42(8), pp.1101-1116.
25. Baggs, D. and Mortensen, N., 2006. *Thermal mass in building design. Environment Design Guide*, pp.1-9.
26. Mahasenan, N., Smith, S. and Humphreys, K., 2003. *The Cement Industry and Global Climate Change: Current and Potential Future Cement Industry CO2 Emissions. In Greenhouse Gas Control Technologies*-6th International Conference (pp. 995-1000).
27. Pade, C. and Guimaraes, M., 2007. The CO2 uptake of concrete in a 100 year perspective. *Cement and Concrete research*, 37(9), pp.1348-1356.
28. Noyce, P. and Crevello, G. (2016). *Durability of Reinforced Concrete*. [online] Structure magazine. Available at: http://www.structuremag.org/?p=9459 [Accessed 16 Dec. 2018].
29. NHBC Standards. (2018). 1st ed. *National House Building Council*, pp. Chapter 9.1 entitled 'External Walls: Tolerances".
30. Black, J. (2017). *DeGarmo's Materials and Processes in Manufacturing*, 12th Edition. New York: Wiley.
31. Wiseman, C. (2000). *Twentieth-century American architecture: The Buildings and their Makers*. New York: Norton. ISBN 978-0-393-32054-1
32. Mortar Industry Association (2013). *Learning Text Part 05: Brick and Block Production*. pp.3 - 5.
33. facts, L. (2018). *LEGO facts*. [online] Nationalgeographic.com.au. Available at: http://www.nationalgeographic.com.au/history/lego-facts.aspx [Accessed 17 Dec. 2018].
34. Thomas, S., Balakrishnan, P. and Sreekala, M.S. eds., 2018. *Fundamental Biomaterials: Ceramics*. Woodhead.
35. National Geographic Society. Wonders of the Ancient World; *National Geographic Atlas of Archaeology*, Norman Hammond, Consultant, Nat'l Geogr. Soc., (Multiple Staff authors), (Nat'l Geogr., R.H.Donnelley & Sons, Willard, OH), 1994, 1999, Reg or Deluxe Ed., 304 pgs. Deluxe ed. photo (pg 248): "Venus, Dolní Věstonice, 24,000 B.C." In section titled: The Potter's Art, pp 246–253.
36. Boaretto, E., Wu, X., Yuan, J., Bar-Yosef, O., Chu, V., Pan, Y., Liu, K., Cohen, D., Jiao, T., Li, S. and Gu, H., 2009. *Radiocarbon dating of charcoal and bone collagen associated with early pottery at Yuchanyan Cave, Hunan Province, China. Proceedings of the National Academy of Sciences*, pp.pnas-0900539106.
37. Cocking, R. and Manufacturers' Confederation, B.G., 2003, January. *The challenge for glass recycling. In Sustainable waste management: proceedings of the international symposium held at the University of Scotland Dundee*, Thomas Telford, London, UK (pp. 73-8).
38. Carter, C.B. and Norton, M.G., 2007. *Ceramic materials: science and engineering*. Springer Science & Business Media.
39. Gailliot, M.P., 1980. "Petrified Lightning" A Discussion of Sand Fulgurites. *Rocks & Minerals*, 55(1), pp.13-17.
40. Dyre, J.C., 2006. Colloquium: The glass transition and elastic models of glass-forming liquids. *Reviews of modern physics*, 78(3), p.953.
41. Pollak, M., Frydman, A. and Ortuño, M. (2013). *The electron glass*. Cambridge: Cambridge Univ. Press.
42. Schroers, J. (2014). *Condensed-matter physics: Glasses made from pure metals*. Nature, 512(7513), p142-3.
43. Fleming, S.J., 1999. *Roman glass: reflections on cultural change*. UPenn Museum of Archaeology.
44. UK Infrared Heating Company. (2017). *Infrared Heating Glass Window | Specialist Heated Glass*. [online] Available at: https://www.infraredcompany.com/pages/infra-red-heating-glass-window [Accessed 19 Dec. 2018].
45. En.wikipedia.org. (2018). *Heated glass*. [online] Available at: https://en.wikipedia.org/wiki/Heated_glass [Accessed 19 Dec. 2018].
46. Worldsteel.org. (2018). *World crude steel output increases by 5.3% in 2017*. [online] Available at: https://www.worldsteel.org/media-centre/press-releases/2018/World-crude-steel-output-increases-by-5.3--in-2017.html [Accessed 19 Dec. 2018].
47. Hans, M., Erbe, A., Mathews, S., Chen, Y., Solioz, M. and Mücklich, F., 2013. *Role of copper oxides in contact killing of bacteria*. Langmuir, 29(52), pp.16160-16166.
48. Oldfield, J.W., 1988. Electrochemical theory of galvanic corrosion. *In Galvanic Corrosion. ASTM International*.pp. 5.

49. Livingston, R.A., 1991. Influence of the environment on the patina of the Statue of Liberty. *Environmental science & technology*, 25(8), pp.1400-1408.

50. Imoa.info. (2018). *IMOA Stainless Solutions*. [online] Available at: https://www.imoa.info/stainless-solutions/archive/36/Galvanic_corrosion.php [Accessed 19 Dec. 2018].

51. Duhl, M.G.D. and Giamei, A.F., 1980. *The development of single crystal superalloy turbine blades. Superalloys.* Warrendale, 41.

52. Graphene.manchester.ac.uk. (2018). *Discovery of graphene - Graphene - The University of Manchester*. [online] Available at: https://www.graphene.manchester.ac.uk/learn/discovery-of-graphene/ [Accessed 20 Dec. 2018].

53. Kharlyngdoh, E. and Barik, S.K., 2008. J. *Bamboo and Rattan*, Vol. 7, Nos. 1&2, pp. 73-90 (2008)© KFRI 2008.

54. Sheltercluster.org. (2018). [online] Available at: https://www.sheltercluster.org/sites/default/files/docs/bamboo_fact_sheet_english.pdf [Accessed 19 Dec. 2018].

55. Guinness World Records. (2018). *Fastest growing plant*. [online] Available at: http://www.guinnessworldrecords.com/world-records/fastest-growing-plant/ [Accessed 19 Dec. 2018].

56. Janssen, J.J., 2000. *Designing and building with bamboo*. Netherlands: International Network for Bamboo and Rattan, pp. 130-133; Van der Lugt, P., Van den Dobbelsteen, A.A.J.F. and Janssen, J.J.A., 2006. An environmental, economic and practical assessment of bamboo as a building material for supporting structures. *Construction and Building Materials*, 20(9), pp.648-656.

57. FAISAL, B. and KINASIH, P., *Experimenting Bamboo as an Architectural and Socio-Cultural Feature. Case Study: The Bamboo House at Eco-Pesantren Daarut Tauhiid*, West Bandung, West-Java, Indonesia. In Arte-Polis 3 International Conference (p. 333).

58. Jayanetti, D.L. and Follett, P.R., 1998. *Bamboo in construction: an introduction* (No. 15). International Network for Bamboo and Rattan, International Development Research Center, Regional Office for South Asia.

59. June 5, 2015, '*Taiheiyo Cement (Japan) Develops New Cementitious Material with Compressive Strength of 400 to 500 N/mm²*' Available at: http://www.taiheiyo-cement.co.jp/english/summary/pdf/150605.pdf [Accessed 19 Oct 2018].

60. Bamboo Import Europe. (2018). *What are the Mechanical Properties of Bamboo?*. [online] Available at: https://www.bambooimport.com/en/blog/what-are-the-mechanical-properties-of-bamboo [Accessed 19 Dec. 2018].

61. Zissis, G. and Kitsinelis, S., 2009. State of art on the science and technology of electrical light sources: from the past to the future. *Journal of Physics D: Applied Physics*, 42(17), p.173001.

62. Zaninovich, S.C., Montti, L.F., Alvarez, M.F. and Gatti, M.G., 2017. Replacing trees by bamboos: Changes from canopy to soil organic carbon storage. *Forest ecology and management*, 400, pp.208-217.

63. https://onlinelibrary.wiley.com/doi/pdf/10.1111/plb.12435

64. Geyer, R., Jambeck, J.R. and Law, K.L., 2017. Production, use, and fate of all plastics ever made. *Science advances*, 3(7), p.e1700782.

65. PlasticsEurope (2017). *Plastics: the Facts 2017*. [online] Brussels: Association of Plastics Manufacturers, p.16. Available at: https://www.plasticseurope.org/application/files/5715/1717/4180/Plastics_the_facts_2017_FINAL_for_website_one_page.pdf [Accessed 20 Dec. 2018].

66. Jambeck, J.R., Geyer, R., Wilcox, C., Siegler, T.R., Perryman, M., Andrady, A., Narayan, R. and Law, K.L., 2015. Plastic waste inputs from land into the ocean. *Science*, 347 (6223), pp.768-771.

67. Niaounakis, M. (2017). *Management of Marine Plastic Debris*. Oxford: William Andrew.

68. Foekema, E.M., De Gruijter, C., Mergia, M.T., van Franeker, J.A., Murk, A.J. and Koelmans, A.A., 2013. Plastic in north sea fish. *Environmental science & technology*, 47(15), pp.8818-8824.

69. Yang, Y., Yang, J., Wu, W.M., Zhao, J., Song, Y., Gao, L., Yang, R. and Jiang, L., 2015. Biodegradation and mineralization of polystyrene by plastic-eating mealworms: Part 1. Chemical and physical characterization and isotopic tests. *Environmental science & technology*, 49(20), pp.12080-12086.

70. Kerber, S., 2012. Analysis of changing residential fire dynamics and its implications on firefighter operational timeframes. *Fire technology*, 48(4), pp.865-891.

71. Braun, D., 2004. Poly (vinyl chloride) on the way from the 19th century to the 21st century. *Journal of Polymer Science Part A: Polymer Chemistry*, 42(3), pp.578-586.

72. Pvc.org. (2018). *History – PVC*. [online] Available at: http://www.pvc.org/en/p/history [Accessed 20 Dec. 2018].

73. Makino, T. (1998). PVC and environmental issues. *Seikei Kakou (a journal of the Japan Society of Polymer Processing)*, 10(1).

74. Epsindustry.org. (2018). *Compressive Strength*. [online] Available at: http://www.epsindustry.org/building-construction/compressive-strength [Accessed 20 Dec. 2018].

75. Mandal, J., Fu, Y., Overvig, A., Jia, M., Sun, K., Shi, N., Zhou, H., Xiao, X., Yu, N. and Yang, Y. (2018). Hierarchically porous polymer coatings for highly efficient passive daytime radiative cooling. *Science*, p.eaat9513.

76. Zhu, M., Song, J., Li, T., Gong, A., Wang, Y., Dai, J., Yao, Y., Luo, W., Henderson, D. and Hu, L., 2016. Highly anisotropic, highly transparent wood composites. *Advanced materials*, 28(26), pp.5181-5187; Li, Y., Fu, Q., Yu, S., Yan, M. and Berglund, L., 2016. Optically transparent wood from a nanoporous cellulosic template: combining functional and structural performance. *Biomacromolecules*, 17(4), pp.1358-1364.

77. Li, T., Song, J., Zhao, X., Yang, Z., Pastel, G., Xu, S., Jia, C., Dai, J., Chen, C., Gong, A. and Jiang, F., 2018. Anisotropic, lightweight, strong, and super thermally insulating nanowood with naturally aligned nanocellulose. *Science advances*, 4(3), p.eaar3724.

78. Song, J., Chen, C., Zhu, S., Zhu, M., Dai, J., Ray, U., Li, Y., Kuang, Y., Li, Y., Quispe, N. and Yao, Y., 2018. Processing bulk natural wood into a high-performance structural material. *Nature*, 554(7691), p.224.

79. Smith, T., Pampanin, S., Fragiacomo, M. and Buchanan, A. (2018). *Design and Construction of Prestressed Timber Buildings for Seismic Areas*. [online] Hdl.handle.net. Available at: http://hdl.handle.net/10092/2631 [Accessed 15 Dec. 2018].

80. Lechtman, H.N. and Hobbs, L.W., 1987. *Roman concrete and the Roman architectural revolution*. Ceramics and civilization, 3, pp.81-128.

81. Jackson, M.D., Mulcahy, S.R., Chen, H., Li, Y., Li, Q., Cappelletti, P. and Wenk, H.R., 2017. Phillipsite and Al-tobermorite mineral cements produced through low-temperature water-rock reactions in Roman marine concrete. *American Mineralogist: Journal of Earth and Planetary Materials*, 102(7), pp.1435-1450.

82. Wiktor, V. and Jonkers, H.M., 2011. Quantification of crack-healing in novel bacteria-based self-healing concrete. *Cement and Concrete Composites*, 33(7), pp.763-770 ; Talaiekhozani, A., Keyvanfar, A., Andalib, R., Samadi, M., Shafaghat, A., Kamyab, H., Majid, M.A., Zin, R.M., Fulazzaky, M.A., Lee, C.T. and Hussin, M.W., 2014. Application of Proteus mirabilis and Proteus vulgaris mixture to design self-healing concrete. *Desalination and Water Treatment*, 52(19-21), pp.3623-3630 ; Wiktor, V. and Jonkers, H.M., 2011. Quantification of crack-healing in novel bacteria-based self-healing concrete. *Cement and Concrete Composites*, 33(7), pp.763-770.

83. Altwair, N.M. and Kabir, S., 2010, March. *Reducing environmental impacts through green concrete technology*. In The 3rd Technology and Innovation for Sustainable Development International Conference (TISD2010) Faculty of Engineering, Khon Kaen University, Thailand.

84. Shoubi, M.V., Barough, A.S. and Amirsoleimani, O., 2013. Assessment of the roles of various cement replacements in achieving the sustainable and high performance concrete. *International Journal of Advances in Engineering & Technology*, 6(1), p.68.

85. Soleimani-Dashtaki, S., Soleimani, S., Wang, Q., Banthia, N. and Ventura, C.E., 2017. Effect of high strain-rates on the tensile constitutive response of Ecofriendly Ductile Cementitious Composite (EDCC). *Procedia engineering*, 210, pp.93-104.

86. Ballari, M.M. and Brouwers, H.J.H., 2013. Full scale demonstration of air-purifying pavement. *Journal of hazardous materials*, 254, pp.406-414 ; Hüsken, G. and Brouwers, H.J.H., 2008, June. *Air purification by cementitious materials: Evaluation of air purifying properties*. In International Conference on Construction and Building Technology, Kuala Lumpur, Malaysia (Vol. 304)

87. Chuah, S., Pan, Z., Sanjayan, J.G., Wang, C.M. and Duan, W.H., 2014. Nano reinforced cement and concrete composites and new perspective from graphene oxide. *Construction and Building Materials*, 73, pp.113-124.

88. Smith, R.A., Kersey, J.R. and Griffiths, P.J., 2003. *The Construction Industry Mass Balance: resource use, wastes and emissions*, TRL Limited, pp. vii ; Khatib, J. ed., 2016. *Sustainability of construction materials*. Woodhead Publishing. Cambridge.

89. Kolesnikov, A.I., Reiter, G.F., Choudhury, N., Prisk, T.R., Mamontov, E., Podlesnyak, A., Ehlers, G., Seel, A.G., Wesolowski, D.J. and Anovitz, L.M., 2016. Quantum tunneling of water in beryl: a new state of the water molecule. *Physical review letters*, 116(16), p.167802.

90. Chipperfield, D., Irace, F. and Nys, R. (2018). *David Chipperfield Architects works 2018, Basilica Palladiana, Vicenza, 12 May – 2 September*. London: Koenig Books Ltd, p.5.

91. Edlin, H. and Forestry, B. (2006). *What wood is that?*. Ammanford: Stobart Davies, pp.30–39.

## 5. Experimenta. Part II:
### The Circle and the Square.
### Thursday

## 5. *Experimenta. Part II*
*The Circle and the Square. Thursday*

### *Reflection, reaction.*
*Scene: Nudging through an active crowd at Portobello market, London. Every few steps, a new aroma. Background sound systems overlap, indifferent to one another, foreground sounds of bustle and noise.*

○ Recently, whilst discussing practices in experimentation, you touched on the idea of 'reaction' space. This notion might be worth exploring a little further that is, how we respond and how a useful 'reaction' might be interpreted.

☐ Yes, it's a perspective that's difficult to define; a more appropriate word might be 'reflection'. Again a suitably abstract analogy might be found in cooking. When we make a meal, or even a cup of tea, we assemble a number of components and ingredients. We mix things together and then sometimes we'll test by tasting. It is in this moment we reflect: 'close but needs more seasoning, or maybe there's too much sauce. How to improve it? Perhaps additional mushrooms to offset the sauce, then more spinach... etc'. It's this internal commentary, when placed on paper, in the context of a developing project, that becomes reflection. The hidden dialogue expressed and edited.

○ But it also reflects a resilience so, even just now for example, someone strayed into our path and you reacted just as quickly, you shifted direction manoeuvred a little and now we're on a new course towards the same destination, and what were you thinking as this was happening?

☐ Exactly. I was thinking this sidestep just revealed the bagel stand!

○ Hopefully readers will recognise the flatness of this analogy. In reality, reflections are composed to be more meaningful, occasionally critical, optimistic, forward-looking, related to other experiences, and hopefully some combination of more than one of the above. It's a marginally more human form of writing, exposing intentions and critical self awareness. This idea of reflection positions the author's thinking side-by-side with the outcomes. It invites a reader to engage in a sense of discussion, as opposed to receiving a monologue. A good ratio of reflection balances out at around a quarter or a third of the text, and appears mostly as the latter part of—or counterbalance to—a conclusion, occasionally in fragments as additional drawing annotation. Ultimately it helps make an already interesting body of work more engaging.

☐ This I can support, but I'm also aware it's *not* the norm. Different schools of Architecture have different aspirations and preferred writing styles. Where do we draw the line between a conventional 'technical' text and a 'reflective' text?

○ Where a course has clear guidelines on the writing style and methods of analysis— as distinct from case studies description and analysis—then of course, attempting something of this nature, to add reflection would be inadvisable. However, where this is supported as an option, it's worth considering ways by which analytical text and reflective text might be differentiated where the two coexist. If the two are arbitrarily mixed, the read will be difficult.

In my experience, this is usually best done typographically, perhaps through a different colour, point size, or font, thus identifying a different graphic layer. Or better still, as a defined paragraph. Allowing the page to be read as it would, the drawing as it would, but then if a reader looks further it's there.

☐ Ah, so, there's a visual hierarchy, a graphic weighting?

○ Yes, in the same manner as a text and its supporting footnotes. In this case, informative text and supporting reflective text. I've also seen it work in the conclusions, where it's consistently the last paragraph of each section, or elsewhere identified by a subheading.

However, for those who've never attempted to write in this manner, especially in the context of something that's traditionally understood to be 'technical' in nature, this might be understandably difficult.

But when it's done well, it can turn a potentially dry read into something engaging and useful.

☐ I've a sense we're maybe also discussing a style of journal writing, something like a travel diary - but without the emphasis on chronology?

○ You may be right, again as a closer form of analogy, it's something of the attitude an explorer might use in describing events, the smaller mistakes made along the way, the tangential thoughts inspired by other new encounters, notes about key decisions taken and why, and so forth, but all the while maintaining an air of professionalism throughout.

It's akin to defining the outline of a puddle, if you ask me again tomorrow the answer might be slightly different.

## Case studies

*Scene: Dense rain outside. RIBA Library inside, hushed tones, background typing. Moving stacks construction section. Calm.*

○ We've been talking about text a lot lately and now we're in the midst of it.

☐ Maybe you can recall your early days as a student, and the trepidation felt on seeing the lengths of each subject's reading list. There are a lot and yet somehow the more you get through the longer it gets. Then, when you feel you're making progress it, the variety of options, materials, technical strategies available tends to be overwhelming.

○ This is just one of the various ways in which we can use a design project as a lens through which to focus options and strategies. Much comes back to that first step, whatever it is, and the idea of some sort of destination. To further alleviate pressure we should see this in the greater context of practice and personal development. The many other projects ahead as the building blocks of competence. Or as it were; the idea of learning.

☐ Here in these book stacks, looking around I'm drawn to the texts that lean on a history of structures, tectonics, and then the more contemporary texts engaging with contemporary sustainable design, experimental construction and computer aided manufacture. It's interesting, to me at least, that many of the books discussing phenomena centred on digital approaches seem to make little reference to the mathematics and ambition of past achievements, from the Gothic to the Roman and beyond?

○ I'd say that's not completely true, but I'd say there's a definite value in learning to appreciate the history of tectonics. For example, here—as Harry Mallgrave points out in the foreword of Kenneth Frampton's *Studies in Tectonic Culture* the culture of architecture has a history of borrowing theory and philosophy from other disciplines for inspiration, yet Architecture has a *'legitimate intellectual development'* of its own.[1] A position echoed more acutely in [Markus] Breitschmid and [Valerio] Olgiati's *Non-referential Architecture*.[2] This history is embedded through the rich layers of thought and interpretation, a representation of socio-cultural practices, and as thesis of tectonic expression; the latter being a reflection of a parallel language of theory and development. The TS is about more than just making things stand up, just as writing is about more than making functioning sentences.

☐ What does this mean for the avant-gardes of digital construction?

○ I think it's difficult to move forward in any meaningful way without knowing you're not perhaps unknowingly repeating patterns of thought, but this time under the guise of an algorithm. I believe there are still many great lessons to be learned from staying open to understanding what has come before, theoretically and tectonically.

An understanding of this nature also demonstrates just how much more effectively we could be using computers, given just how far things have been pushed. Just off-hand, I'm thinking of examples such as King's College Chapel in Cambridge *(Fig.1_2-2)* and the engineering accomplishments of its fan-vault masonry ceiling.

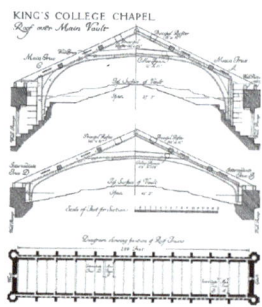

*fig.5_1-2: Fan Vaulted ceiling: Reginald Ely, Simon Clerk and John Wastell, Kings College Chapel, 1446*
Weighing nearly 2000 tonnes the very thin stone ceiling spans 12.66m and varies in thickness between 100–150mm.

☐ True but elsewhere in computing and construction we have seen improvements in other aspects of practice such as; time-cost improvements, a greater understanding of our impact on the climate, coupled with more precise environmental management. On the inside sleeve of an older edition of Reyner Banham's *The Architecture of the Well Tempered Environment* it is remarked that, *"As much as half the cost of and even more of the design effort in modern buildings goes into mechanical devices for communication, lighting, ventilation, heating and environmental control generally"*. This is not proportionally reflected in the documentation of the history of modern architecture. The notion of *"modern architecture as a complete art of environmental control, combining structural and mechanical methods"* is also mentioned.[3] The remainder of this book makes a good effort to position these developments in a human and historical context, but I would agree, there's a great deal to be learnt from past masters.

◯ Also, there's more to be gained from first hand investigation. Visiting buildings of interest and creating a photographic collection of both failing and successful details, reaching your own conclusions and finding new case studies in the process.

☐ In the manner of Thomas Coryates and the early English aristocratic scholars of the Grand Tour? [4]

◯ In a way, it's rare and it's great to see self developed investigations and the thoughts they have inspired from and amongst imported case studies, but only where there are useful examples, and within easy reach.

And let's not forget the non-architectural, non-obvious case studies or precedents. Examples might be found in nature, installation art, design products and other parallel disciplines. In my view it's important to establish common questions before starting: Which part of this case study am I interested in? Beyond appearances how else might this case study be relevant to my TS? Are there any lessons to be learned—things not to do? How can I now improve on or further develop this idea?

For improved readability, the questions are then omitted from the developed text.

As a last tip, in the communication of a case-study: where a study presents its

---

1. Frampton, K. and Cava, J. (2007). ***Studies in tectonic culture***. Chicago, IL: Graham Foundation. p ix.
2. Olgiati, V. and Breitschmid, M. (2018). ***Non-referential architecture***. Basel: Simonett & Baer, p.13.
3. Banham, R. (1984). ***The Architecture of the Well-Tempered Environment***. University Of Chicago Press. pp Jacket.
4. Moore, T. (2002). ***Continental drifter—taking the low road with the first grand tourist***. London: Abacus.

own interpretations of diagrams, in its own graphic style, with its own system of annotation—as relevant to the interests, it serves to demonstrate understanding.

## *Document formation.*

*Scene: On the London Underground, switching lines at Westminster. An interior built like an exterior, or two mice in a steam-engine.*

○ So far we've mused on just a few of the facets of a TS, some perhaps more effectively than others. Are there any thoughts on how to go about bringing these sides together to build a more complete entity?

☐ This idea of document structure, and its presence are central to the portfolio. It speaks to how the TS as a body of work is perceived in context and then how it's put forward.

When we studied it was a document we could easily read with—or happily without—the main portfolio. We had a set of key criteria to present and there was flexibility as to how we chose to narrate or layer it. This method has largely endured, with an albeit less theatrical approach to the 1:1 exercises.

○ I also like to encourage my students to enrich the quality of individual elements, especially the diagrams and photography. There's a lot to be said for a consistent flow of clearly documented model photographs, experiment processes imagery, 1:1 sample scans and so forth, and there's always room for typographic experimentation within the diagram. I think, a good diagram often looks deceptively simple. Diagrams support drawings, concept analysis, structural grids and studies, environmental strategies and sub strategies etc.

It's important that these graphics, these photographs speak of something significant about the project, a driving reason for their direction.

☐ How do you mean?

○ Ah, as a small example, I was looking at an interior view, a well constructed photograph of a large scale model of the back of a living room space with beautifully captured light. This particular proposal used a reflective balcony surface to bounce diffuse daylight onto the otherwise darker ceiling to the rear of a deep plan interior.

This framed view contained the idea of a lighting strategy. Accompanying this image was a smaller diagram section showing clearly the idea of the light reflecting and a typical arc of daylight hours. Further on there was a concise exploration of robust but also reflective and anti-slip material options, all of which contributed to an over-arching interior environmental strategy.

☐ Interesting, but what of the larger principle diagrammatic sections, addressing heating and ventilation strategies?

◯ This methodology was present throughout the document, it spoke of a more, and I hesitate to say, phenomenological approach towards expressing technical strategies and principles. The desired experience of interior space took a higher priority, and the remaining strategies and studies were positioned to support this idea. The larger macro strategies were also present, they just weren't dominant.

This careful balance - between a representation of technical frameworks and the idea of an intended experience - is largely determined by the design studio's orientations and individual preferences.

☐ Balance, as a target, sounds appropriate. I don't think the two sides: technical method and spatial intention, should be thought of as mutually exclusive and I suspect the hierarchical difference between the two - at least as they sit on the page - doesn't necessarily need to be so separate?

◯ Ideally, but I can say that our bias is for the experiential aspects of a design proposal, which I think reflect the nature of how a TS is presented - enmeshed within the main portfolio. In both cases however, an understanding of tectonics should be strong. The time taken to explore technical possibilities and the subsequent decision making process should be equally strong.

☐ Noted. Quietly retreating.

◯ An individual's narrative structure needs to be defined and easily followed, so within this there remains room for how and where the TS criteria are met. This idea of quality does need to permeate through the way in which case studies are presented and dissected; texts and other books are effectively referenced; diagrams and photographs labelled; experiments documented; drawings scaled; line weights distinguished; text proofed and so forth.

These should all work hand-in-hand with the core design project's representation, and with design scheme intentions, all properly expressed as the key rationale for tectonic decisions. Plus, if this isn't a long enough list, the whole thing needs to feel motivated. Principally, a text that was interesting to write is usually also interesting to read, and the same I believe is true of the opposite!

☐ A robust list, worth pasting up beside every studio desk! And I concur, these are the little practices that eventually combine to build the feel of a serious document. At first glance I think it could easily feel like a daunting list of points, but when taken in bite size pieces, through multiple passes, any document will start to feel like a whole, fairly quickly. It's important to be able to pick up a TS and understand everything about its design proposal, and then everything about its architecture.

○ Absolutely, and returning briefly to the usefulness of 'motivation', we haven't really dug into the value of a good research question, as the basis of a TS direction, and touch point for later reflection.

☐ We could easily have several more discussions about the shape and nature of an attractive research query, it's something I've often struggled to describe—in the sense of: how best to define an effective question?
   In my estimation, the more interesting studies are inspired by focused curiosity. As such, recognising the technical dimensions of the main design project's own interests helps to translate such curiosity into a opening question—or questions. I think, by inviting more informed investigations we might stimulate even more interesting research and hopefully a more in-depth understanding of the technical options at hand.

○ However, the simple questions are also always worth asking. I've often seen these lead to unexpectedly useful positions, provide launch points for subsequent experiments, or help to frame key subject areas. A prerequisite to everything we've discussed. In this manner the question presents a sort of opportunity, or at least a potential starting point to help focus the scope and composition of the document as a whole.

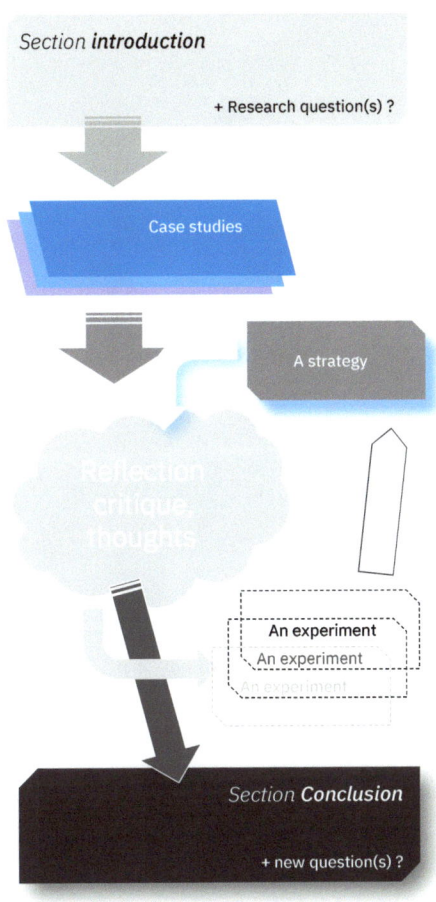

*fig.5_3: An example process for structuring a chapter. Reflection and experimentation as an iterated process*

*6. Approaching prototyping*

# 6. *Approaching prototyping and the idea of the 1:1: a tangible component*

Here we dwell momentarily on the idea of prototyping and its role as a nested sub-process—within a larger design process. The idea of the prototype as a means to develop the tectonic language of a project, and within itself; prototyping as a valid form of experimentation.

## *Approaches to making*

The process of making samples - to learn from, to experiment with, to demonstrate through -goes back to beyond the seventeenth century and the days of the Grand Tour, a northern European tradition that was one of the earlier forms of architectural education. In the UK this journey of learning, taking aristocratic students across Europe on a cultural pilgrimage to Rome, sometimes culminated in the construction of garden follies upon their return. These follies are large objects of curiosity that sometimes served as built collages of architectural motifs studied overseas. The follies demonstrated newly found technically acculturated knowledge.

Thus, the investigative approaches of learning from case studies, reading buildings, and building prototypes reflects one of the of the earliest traditions in architectural education. A way of seeing which made little to no distinction between the art and the craft, a division often attributed to the modernist tradition.

Prototyping is itself part of a process of reiteration or cyclical development. Prototypes are beyond the essence of the working model in as much as they typically aim to employ construction materials and operate at the 1:1 scale. The tacit perception gained from hands-on interaction with materials, through physical interrogation of their qualities, in many cases leads towards a deeper understanding that cannot be conveyed through words alone. That understanding occurs in terms of a deeper sense of 'insight' and rather than 'knowledge'. [1]

Richard Sennett's *The Craftsman* briefly discusses the discovery process behind the Bilbao Guggenheim facade panel design and how, for the team involved, trialling methods for manipulating titanium panelling elicited a new understanding of structure.[2]

1. Hasegawa, G., Siza, A., Olgiati, V., Märkli, P., Lacaton, A., Vassal, J., Flammer, P., Geers, K. and Van Severen, D. (2015). *Go Hasegawa, conversations with European architects*. Tokyo: Lixil, p.103.
2. Sennett, R. (2008). ***Craftsman, The***. New Haven: Yale University Press. ch.8, pp.222-225.

Here 'discovery' is a keyword. It offers an open approach to prototyping to help create a state where critically new ways of thinking are discoverable.

As such, sometimes you do the thing first and the (new) inspiration comes after. Even where the process is linear, it rarely is always one way, going forward. Therefore, instead of viewing prototyping as a one-step task to make a final piece, it is better considered part of a continuing process of experimentation.

As so much of the act of 'building' is delegated, as in many cases for an architect, the making of a prototype is principally the closest we get to the hands on task making of a building. Ultimately during the design stages, the fabrication of a prototype reflects the first moment where a proposition and reality come into contact with one another. The key objectives and ambitions of a project are brought into existence. As such, it is worth determining which situations within the framework of a proposal are likely to become the most representative of the scheme's intentions. This might be anything from a built detail indicating a typical relationship between common materials, to something more akin to an installation piece simulating a momentary experience from a lighting or other environmental strategy.

*fig.6_1: Frank Gehry, Guggenheim Museum, Bilbao. Refining an understanding of structural behaviour through prototyping. The resulting facade system came about through rigorous experimentation and prototyping*

Other benefits of prototyping include making it easier to picture the construction process by which the project might be built i.e. the build sequence. This helps to refine and visualise the tectonic language of a project at a more tangible scale.

Example prototype formats might include:
- Something as small as a material sample, an experimental form of cladding, interior panel system, window framing approach, new type of sliding door mechanism, responsive method for light diffusion etc.;
- Transformative mechanisms including reactive facade systems, window pane opening automation, mechanised ventilation panels, heat dispersal system, auto positioning solar shade systems, which in turn might be hydraulically operated, manual mechanically, digitally automated etc.;
- Background fixing systems or techniques addressing assembly including panel clipping connector systems, locking mechanisms, adhesive processes, new casting techniques, folding processes, and combinations thereof etc.;
- Environmental/experiential such as room sized installations - where heating lighting humidity airflow and other qualities are manipulated to represent an intended moment or experience in the design project and;
- Something as large as a built piece of a facade detail exploring a promising solution where two or more building elements (floors, walls, doors, windows, roof, ground etc.) converge.

There is no intended hierarchy to this list, and it is only partial. The key principles to keep in mind are that there is always room to experiment with more than one prototype, and the chosen format(s) should correlate to the nature and direction of the main design project.

## *Recipes for making recipes*

The process by which experiments and prototypes are developed should be replicable. This means it should be possible for another person to follow the same process recorded for the experiment and arrive at the same outcome. When exploring a new idea, keeping a record of decisions and progress including material, quantities, dimensions and timings, where appropriate, is to leave a trail of bread-crumbs. Being methodical in documenting your process enables others and yourself to retrace the path taken and thus reproduce or augment the same results.

Technology transfer is a term used to describe the movement of ideas and discoveries across different industries. For example, the development of windshield technology which eventually found its way into more robust touchscreen phones and tablets. The idea of technology transfer is itself a useful and potentially more interesting means of generating new ideas or directions for experimentation. For example, if an architectural problem calls for the movement of large facade sections or walls, then alternative solutions might be found in aircraft hanger detailing, or theatre rigging design where systems have been developed for moving large and heavy backdrops quietly, quickly and precisely. This opens the view to other disciplines and other forms of practice—from fashion to product design to …? Perhaps a question of creativity in observation and translation.

Looking within our own digital and manual processes—of modelling and casting—we find projects in architecture inspired by physical interaction with tangible materials. Common and uncommon experiments such as Christian Kerez's *(Fig.6_5–8)* 2016 Venice Biennale, where some 300 prototype samples were developed in various material mixes through the search for a preferred texture and form. Or Studio Mumbai's *(Fig.6_4)* general integration of full scale carpentry mock-ups, a process of building 1:1 elements as a counterbalance to detail drawing development.

An interest in both the developed object, and (separately) the process of development, as parallel studies. A development of both is ideal, however, one may be enough if it is in itself considered suitably interesting. That is to say that a new form of prototyped tile might use an otherwise normal process of tile making, but perhaps one of the ingredients is different, or a slightly different mix that in itself is nothing special, but yet the outcome or form becomes the focal point of attention. Or vice versa for the process; where the example tile doesn't in itself appear to be particularly interesting but the process behind its formation is seen to be considerably different to other more normal recipes for manufacturing similar

*fig.6_2: Prototyping at scale in practice: Shake table testing earthquake resistant structures*

forms of tile. Perhaps the process leads to significant energy savings, or reduced production of $CO_2$ or significant savings in time, or increased compatibility with other processes used elsewhere in the same project, and so forth. Process and product as two parallel avenues of discovery, the story of the journey as one thing, and the artefact of the destination as another.

*'Perhaps the path we have traced ends logically in the poet William Carlos Williams's declaration in the 1930s that there should be "no ideas but in things." The poet was sick of soul talk; better to dwell in "things touched by the hands during the day." This has been the craftsman's credo in the past.'* [3]

*fig.6_3: Catenary structure study of Antoni Gaudí's (1852 - 1926) Sagrada la familia, Barcelona*

Using an elaborate network of small bags of sand hung to form curvatures which, when viewed upside down via a floor tilted mirror, enabled Gaudi to approximate the angles required for the structure to work.

3. Sennett, R. (2008). *Craftsman, The*. New Haven: Yale University Press. ch.4, pp.146.; Williams, W. (1970). *Imaginations*. New York: New Directions Pub. Corp, p.110.

The Workshop, Mumbai

The office of Studio Mumbai Architects incorporates a large proportion of skilled carpenters working alongside the designers, through an integrated studio and workshop. This form of practice provides the ability to both design and test the implications of large components and building segments first-hand. Operating in this way affords a more direct understanding of material properties at scale, and useful appreciation of the spatial effects resulting from design decisions.

fig.6_4: [Opposite and above] Workshop space featuring prototype façade elements and other works

## Incidental Space, Venice Biennale 2016

Aspiring to represent nothing other than itself the Swiss Pavilion by Christian Kerez simultaneously embodies a tectonic and spatial exploration, developed in part to invoke questions about the production and experience of architecture.

The design approach was momentarily inspired the incidental interior spaces formed in and amongst a sample of office 'garbage'. Seemingly incidental, the following process involved rigorous experimentation and trialling of physical prototypes resulting in over 300 variations. After careful evaluation, prototype number 180 was chosen. The prototypes were themselves experiments in material formation, the original batch was developed from (amongst other materials) sugar and dust cast in plaster, leading to different forms with their own inherent

ornamentation and surface qualities. The following digital modelling and rendering (including VR) processes were determined by the physical prototype after being dissected and scanned. The final form was derived through a combination of foam moulds formed by hand, by digital 5-axis milling, and other more complex parts by 3D printing. Part of the structural strategy makes use of the increased and variegated surface area and the inherent qualities of sprayed fibre cement at 20mm thick. For the final installation; craftsmanship was employed through the construction skills of a Hollywood film set fabricator - who happened to specialise in the fabrication of large scale transportable geological scenery. Peter Zumthor reflected on this as an interesting example; combining craftsmanship 'the hand' and digital processes. In this instance the final installation was itself a 'prototype'.

Overall this project demonstrated the possibilities in designing and seeing through an otherwise untested process, centered around material experimentation and approached with a spirit of naive openness–as described by Christian Kerez–and advocated by the office to students addressing similar prospects.

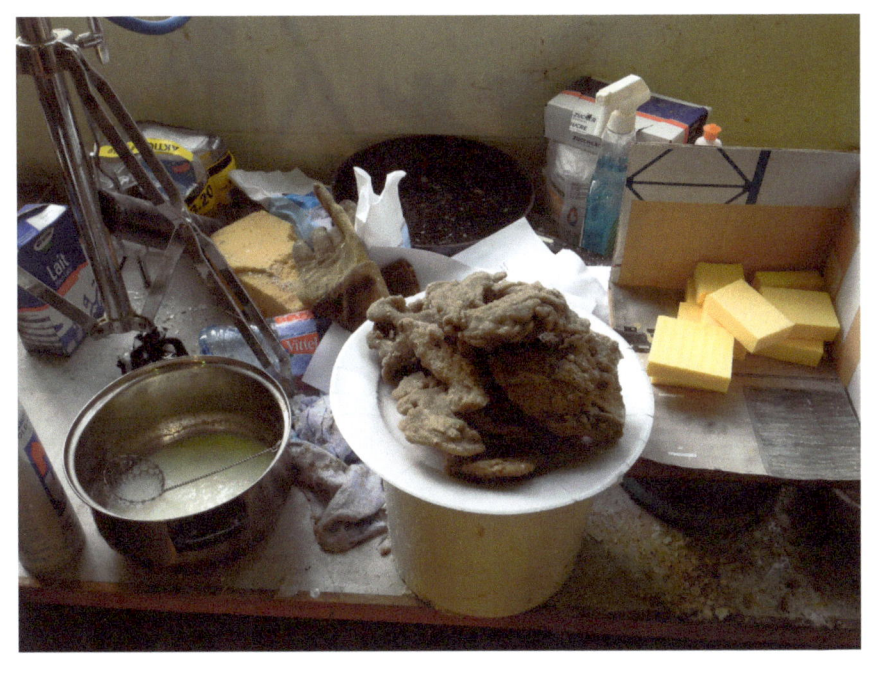

fig.6_5: [Opposite] Prototyping

fig.6_6: Prototyping: Various material mixes to develop the positive before creating the mould

*fig.6_7: Prototype #180 dismantled for 3D digital mapping*
*fig.6_8: [Opposite] Venice Biennale installation*

# 7. Next level

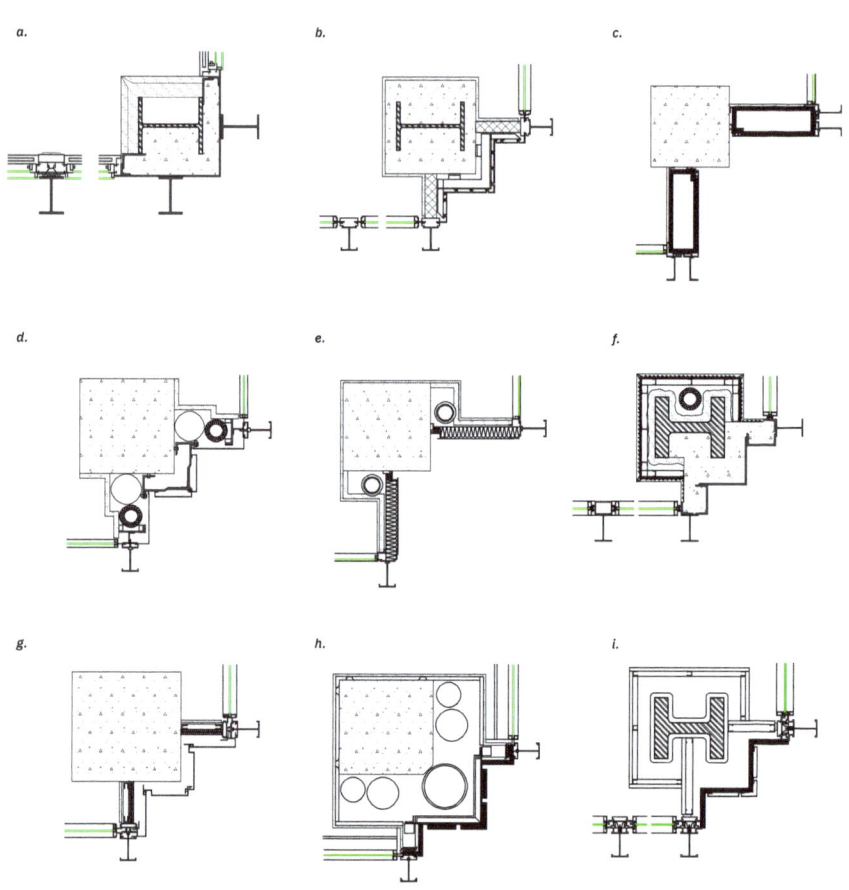

fig.7_1: Mies van der Rohe, corner study. 9 of 15 Redrawn from an original study by John Winter [1]. Tracing a tectonic movement from Heck's illustration of classical the orders of Fig: 1_2

| | | |
|---|---|---|
| a. 860-880 Lake Shore Drive, Chicago, 1951 | b. Seagram Building, New York, 1958 | c. Pavilion Apartments, Lafayette Park, Detroit, 1958 |
| d. One Charles Center, Baltimore, 1963 | e. Lafayette Towers, Lafayette Park, Detroit, 1963 | f. Federal Office Building, Chicago, 1964 |
| g. Westmount Square, Montreal, 1968 | h. East Wacker Drive office building, Chicago, 1970 | i. IBM office building, Chicago, 1971 |

# 7. *Next level*

## *Moving forward*

New aspirations tend to follow new abilities. For those with prior experience of having produced a Technical Study (TS), greater depth, more ambitious experiments, or further analysis should be considered a means to raise standards in tectonic practice as both a craft and an art. In the spirit of pushing things further, this chapter aims to explore, various notions of what a more ambitious TS might choose to incorporate.

Just as for chapter 1: 'A Preamble', many areas remain untouched including designing for fire, opportunities in building regulations, building accessibility, notions of sustainability etc. These will be left to the scope of other texts, TS tutors, the discretion of the school, and the skill and judgement of you, the reader, to recognise and incorporate.

Instead, here we find a range notes addressing:
- web based technical analytical methods for testing propositions;
- a contemplation on the mass of a building;
- further ways of thinking about structure;
- brief thoughts on sensory perception in environmental design;
- ever evolving ideas of technology in construction;
- and the inherent values of robust detailing.

## *Means of modelling*

Concerning the methods available for designing modelling and analysis, recent developments in internet browser software and Cloud computing have opened up various website based applications (Web Apps). This approach to software allows for a potentially high level analysis of structural deformation, thermal behaviour, airflow, sun and daylight modelling through the browser, independent of other desktop applications *(Fig.7_2-10)*. As such Web Apps are useful in the absence of dedicated desktop applications or, in rare cases, where substantial Cloud based computing power is required. A handful of examples with site addresses, found at time of writing, are included on the following pages. Internet searches should reveal further, potentially more suitable Web Apps, depending on the nature of the diagram or scenario required.

As a rule, diagrams (or other App based outputs) should *only* be used in a TS if:

a. There is an interesting discovery of some manner—whether positive or negative—followed by a response or further discussion;
b. The diagram or App based output is annotated—ideally in a way that also explains its presence/role in the study, and with units where relevant;
c. The diagram or App based output reinforces or demonstrates some form of design resolution at a given stage of development, or a particular project detail.

These points aim to avoid disassociated graphic components. More than the 'appearance' of being interesting, diagrams should aim to explain an idea and relate to the TS narrative.

## *Combined matter*

Here we are referring to the 'mass' of architecture and more specifically the 'build-up' of a wall, a floor, a roof etc, the resultant layering of materials of differing properties and roles, and whether there might be advantages in considering the one (or fewer) materials—a stereotomic attitude, or the many combined materials—a tectonic attitude.

This idea of 'combining' has itself often been considered by many as a way of assessing tectonics in construction, from monolithic structural form to the layered component based form. When one element of a building becomes many sub-components, and then when each sub component develops, we have not just a subdivision of elements but the specialisation of each sub-component. By way of example; when what would have been a solid stone wall is remade as a brick cavity wall, the idea of what was a singular mass is reconsidered as a series of layers—internal lining, structure, insulation, vapour control, weatherproofing. Each of these layers is then allowed to evolve as a different specialist material. Stereotomic henceforth refers almost exclusively to concrete, or wood, or cross-laminated-timber, or stone, or rammed earth, etc. i.e. a constrained material palette where the elements of a building are considered to be made from a single material. Combined, henceforth, refers to the layered nature of cavity wall construction or 'Hi-Tech'—where for example; steel frame construction is further clad in a variety of differing material panel options and sub-components.

Many argue for a notional return to building the one prime material, as a means to focus architectural intentions or the idea of a project, though in theory there may be just as many ideas better expressed through a vocabulary of materials. These considerations, in relation to the architectural intentions of design project, are worth reflecting on. Recognising and playing to the advantages of the most suitable direction may help to reinforce the thesis of TS.

| | + | - |
|---|---|---|
| **One / stereotomic / monolithic** | Monolithic allows for a clearer tectonic understanding of an architecture. Pragmatic, simpler to maintain (in principle) potentially longer lifespan, simpler details, clearer more environmentally accountable industry supply chain. | Typically, this leads to thermal mass and as such is not ideal where a rapid heating or cooling response is desired. Reduced palette of suitable materials, structurally limited in height, if wood rainscreen (a second material) is typically required, service runs may be harder to conceal, if necessary. |
| **Many / tectonic / combined** | Separation allows for more granular control, adaptation for different programmes and functional requirements, between spaces and between inside and out. Able to maximise the benefits of more materials, such as lightweight insulation for rapid thermal response, and metal to build taller. | Other drawbacks include significantly increased complexity leading to more expensive details/ detailing, combined materials have different lifespans which in turn can compromise general robustness, potential interstitial issues such as combining materials invariably leads to different methods for attachment, mixed materials can lead to a more confused architectonic thesis. |

*Table.7_1: One and many*

This over simplified table (Table. 7_1) is in itself a perspective, an overly casual means of categorising construction attitudes and building typologies. However as suggested, awareness of this, as for other perspectives, should allow a project to better define a 'position' or an attitude in a more purposeful manner.

## *Statics and haptics*

There is a difference between the structural properties of a building element and the feeling of structural security offered by the same element. This difference might be something intentionally played on by a project, or not, depending on its what? nature. This difference is felt through the haptics, the feeling of response of, for instance, the tiny amount of movement felt from a typical glass balustrade if pushed reasonably hard, perhaps enough to trigger a sense of instability - even though it's still safe. In actuality it might well be technically capable of supporting the weight of many without failing. It is still safe, just not as physically reassuring as say a thicker concrete balustrade. In some circumstances this might be considered less attractive than the thinner glass option, it triggers a sense of security through uncompromising rigidity and a feeling the more robust balustrade could easily support the weight of many more. These issues should be considered with regards to the tactile sensation of a structure or the *quality* of the real. This sense of tactility often accompanies an intended perception of a structure.

So here in this difference, there is an opportunity for technical reflection and the possibility of explorations on the technical distance between structural capability including fire resistance, and structural stability. Here a stable structure becomes an interesting variable: at one extreme unstable structure addresses the sense of catastrophic collapse as a direct feeling of risk, but at the other end, lesser or smaller movements might just impart the feeling of a lower quality but stable form of construction. Given these parameters and the design intention of a project, how do you determine the ideal perceived and experienced thickness of a structure?

After addressing the primary structural design factors, including foundation system options and implications, the more advanced TS's engaging in this area often ask questions of this nature. Those questions involve thoughts on movement and thoughts on the haptic experience, considering how to unite the two.

## *Environmental Qualities, Environmental Contexts*

When exploring intentions for a sensory experience of a project and the resulting environmental factors, consider the qualities available within each phenomena. To name just a few in passing, we might consider:

- the many kinds of light—ranging from the cold (blue) to the warm (orange);
- the speeds of heat—ranging from the instant (electric) to the slow (water);
- the feel of acoustics—from the echoing reverb of an institutional space to the muted still air of a recording studio.

Including the additional senses that enhance our perception:

- of touch—and the few places we actually touch buildings despite their size;
- of smell—the olfactory perception of materials and the control of ventilation systems for effect;
- of taste—it's inextricable connection to smell;

to the many subtler forms of perception that exist between and around these.

Through meditation on the points above we may further explore the plans and sections of a proposals environmental landscape. Are there areas intended to be cooler or hotter than usual? Given the lifestyles and usage patterns of the project how might different speeds of heating impact the way we interact with it? Of the different types of spatial intentions would the application of different temperatures of light in different interiors improve the design position? How might we direct or play with airflow or humidity to demarcate special areas? Or how might possible combinations of these factors be of use?

Although this list is short, it reflects one way of seeing and thinking about spatial environmental qualities and an approach that recognises the possibilities available if we look beyond the more obvious effects of lighting, heating, sound and so on.

Thinking in terms of the macro, larger scale environmental design decisions extend beyond ideas of the interior experience. These decisions will touch on greater global concerns of resource consumption, responses to climate change, land use and sustainability. At the scale of the building we may consider energy use, air and water management as design factors affecting heating strategies health and plumbing services respectively, all backed by a body of interesting case studies from which to garner inspiration. From here, expanding beyond the building envelope, further speculation should consider current research, to imagine larger scale design strategies and their potential impact on the more pressing environmental concerns.

Given the forecasts of global environmental change; how do we go about making proactive ecological design decisions—whilst retaining other design agendas?

Many texts addressing low energy environmentally sustainable design are assumed to propose (or even impose) practices with stylistic design implications. Many of these practices are quite flexible and may be adapted without affecting performance outcomes. However, to achieve these outcomes—whilst maintaining a design agenda—requires a holistic and more complete understanding of building physics, material characteristics, and construction processes. For these reasons, as much as sustainable low-energy design thinking is encouraged, ongoing discussion with design and TS tutors will become essential to balance and better integrate appropriate agendas. Discussion should also help to identify technical exceptions that may occur where rules are combined and applied within a specific project scenario. In addition to the applicable guidance, there are many other factors which may need to be assessed elsewhere as they extend beyond the traditional scope of a TS; into social and political practices.

In maintaining focus, all considerations should be lead as appropriate by the interests of the core design project. Hence it may be advisable to start with the more relevant factors, using the remainder as discussion points, before expanding outwards to see where the research leads.

Ideally, various relevant ecological strategies would intertwine with respective subject areas of a TS, and not be objectified as an individual chapter—though both (integrated and separate chapter) is also possible.

## *Tech and technique*

Expanding our inquiry—and where appropriate the narrative of a design project thesis—to speculate on future directions in the construction industry may invite new forms of investigation. These new questions might address developments such: as the impact of robots in construction; advanced compression shells and Rubens structures[2]; carbon positive design; maverick approaches to structural design; and/or other interesting technologies emerging in building physics. However, do not be disheartened if research in these directions does not develop as intended. In professional practice, many of these subjects are considered '*state of the art*' (at the time of writing this book) and as such there are relatively few offices themselves able to devote sufficient resources to understanding and applying these in practice.

In tandem, further consideration or thoughts on exploring the 'meanings of things'; ontological understandings of materials and the idea of detail composition, may serve to help balance the hyper-technological. As discussed further in K. Frampton's *Studies in Tectonic Culture*[3] and M. Cadwell's *Strange Details*[4], the exact language of assembly of a detail can express the position or attitude of a project on a sliding scale between handmade and machine-made. Other associations might include low-budget and high-cost, or ephemeral and tactile, and so forth. This level of tectonic range resembles the graduated difference between learning to play an instrument, and creating compositions with the same instrument. Aiming to operate at this higher level should help to demonstrate an understanding of not just the materials in play, but the many combinations of options available in connecting them to one another. Ultimately, these efforts aim to demonstrate both coherence in detailing, and coherence in idea.

## *Detail response*

The practice of architecture has a long history of detailing, one of which the further back in time we travel the lower the (perceived) complexity, the fewer the number of combined materials. Within this same history we can find a greater tacit knowledge of these materials, a wisdom of stone, brick, timber developed over generations understanding the ways of weathering and the methods for treating these materials to achieve enduring forms. Today and in the future we have more material options to contend with, many of which haven't in themselves existed for the time spans we would hope to build for. This range is expanded in part to the increased performance requirements and expectations we now place on architecture. Tallying these options and practical needs tends to lead to more complex details. Added to this we may need to consider an equally disparate spread of manual labour skills—though not yet factoring robotics. This is of course in many ways a generalisation, but it is common for complex details to be developed, with equally difficult (sometimes impossible) assembly sequences, handed to crafts people of differing abilities, or installed in high impact situations subjected to extreme wear and tear. We should not be surprised when or if these details fail.

However, as our experience with detailing increases, and as we become more comfortable with material strengths and critical weaknesses, we find new heights to reach for. This is to say; as we gain confidence in a new game it's natural to raise standards and recognise new targets. Such targets, as hinted above, might include buildability or expression beyond function or robustness. Buildability largely depends on experience accrued through studying the order in which things are placed together, and then how things tend to be put together. This comes through observation, consideration, and conversations with crafts-people of various specialisms, and again with workers of no specialist ability. Robustness is developed through studying the innate qualities of materials, notions of how far they can be pushed, how they fail, their availability for the given locale, best practices for maintenance, common causes of decay, heat or moisture related movement, and typical material life-spans given their intended conditions. There are many starting options, but one of the easiest—in most cases—is to start with the material itself, preferably with a sample or two in hand, before building a narrative through investigation. How available is this material in this locale? What is a typical usage life-span for this material? What special fire precautions are required? What level of environmental impact does what carry? And how can this impact be reduced? How much of a given project or individual space is anticipated to be of this quality and texture? Which other materials would complement the whole?

How will this interface with the structure? And what is this structure made of? And so on. Many questions can emerge from just one viewpoint, but familiarity with these approaches is in itself a longer term aspiration. In most cases, these go beyond the scope of a TS and are be taken in small steps in due time.

Richard Sennett's *The Craftsman*[5] paints a perspective of craft and its role in society. There is no doubt that 'on the ground' skills are needed to bring a building into form but we can't always rely on it being available in the right places and at the times required. Instead, another approach is to assume otherwise and design for easy on site construction and here, by 'construction', we mean 'assembly'. This might mean finding and weaving together different building product systems from different sources, whilst taking the time to find further standardised or robustly manufactured interfacing components to allow for secure connections—and here is where the detail occurs. Between the craftsman and off-the-shelf systems we have numerous detail design attitudes representing differing positions. Once a direction is established we then need to appreciate the skill-sets available and possible alongside what is ultimately desired for the project in question.

## *Coherence*

By nature, tectonics today is a diffuse field of study. As such, the many varied points highlighted throughout this book should be taken as introductions, each worth deeper investigation.

Maintaining an overview within your TS, with a sense of narrative or a clear sense of destination, will present its own set of challenges. To holistically shepherd together the various, occasionally disparate, forms of investigation under the one umbrella is in itself a form of achievement.

Do not be surprised if your study takes you to unexpected places, does not turn out exactly as planned, or shifts direction, this is the better nature of exploration. How well these new outcomes are integrated into an adjusted narrative is an additional form of achievement. Depending on your own course schedules and design situation—these discoveries should be allowed to contribute to your design project as and where appropriate.

When starting out, to avoid becoming overwhelmed it is important to resist the pressure to tackle *everything at the same time*, generally speaking: peripheral interests will have their turn in future projects. Where uncertainty remains, recall your initial TS questions and do not be afraid to fully explore the possibilities of the one material, the one structural strategy, and the one environmental design principle. This approach will help to understand the one case-study, and engage more fully with the first experiment. From this position the second material, the second strategy, a second additional approach etc will follow-on more constructively. Leading to the third and so forth, to wander with intent until the TS finds its thesis.

Beginning, middle, end: introductions; processes; conclusions. *(Fig.5_3)*

Finally, the reference texts of chapter 2: 'The Field' will illuminate the subject elements, whilst raising the level of conversation with your TS and studio tutors.

This book has aimed to encourage further tectonic thinking and the composition of a meaningful series of explorations, all to inform your TS document narrative. Between a design project's principle interests and the materials at hand which question(s) would you like to begin with?

1. Winter, J. (1972). The Measure Of Mies. Architectural Review, 153(900), pp.104–105.
2. Retsin, G. (2019). *Discrete: Reappraising the Digital in Architecture.* [S.l.]: John Wiley & Sons Austral, pp.54–61.
3. Frampton, K. and Cava, J. (2007). *Studies in tectonic culture.* Chicago, IL: Graham Foundation.
4. Cadwell, M. (2007). *Strange details.* Cambridge, Mass.: MIT Press.
5. Sennett, R. (2008). *Craftsman, The.* New Haven: Yale University Press.

Web Applications

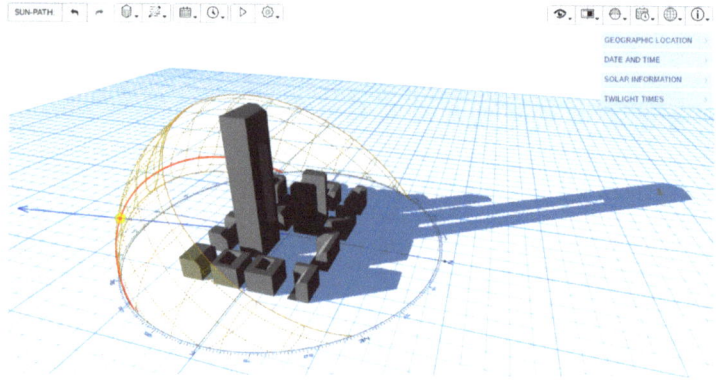

Sample output from a site that produces geo-positioned animated sun-path studies.
http://andrewmarsh.com/apps/staging/sunpath3d.html

*fig.7_2: Sunpath 3D—Solar path modelling web-application*

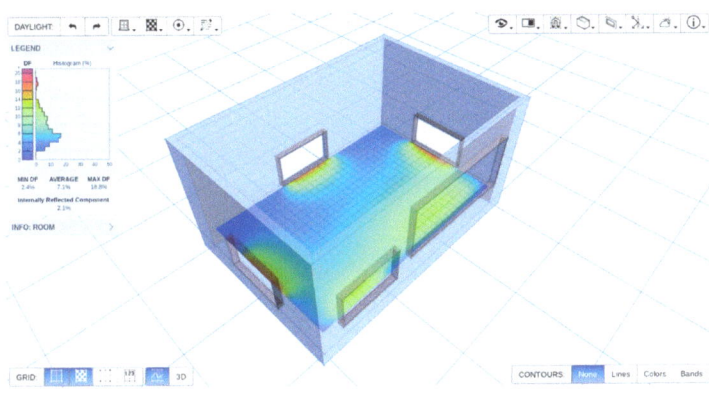

A simple method for assessing interior daylight spread. Easy to interact with though restricted to a cuboid room format—but of any proportion or number of windows.   http://andrewmarsh.com/apps/staging/daylight-box.html

*fig.7_3: Daylight Box—Interior daylight modelling web-app*

# Web Applications

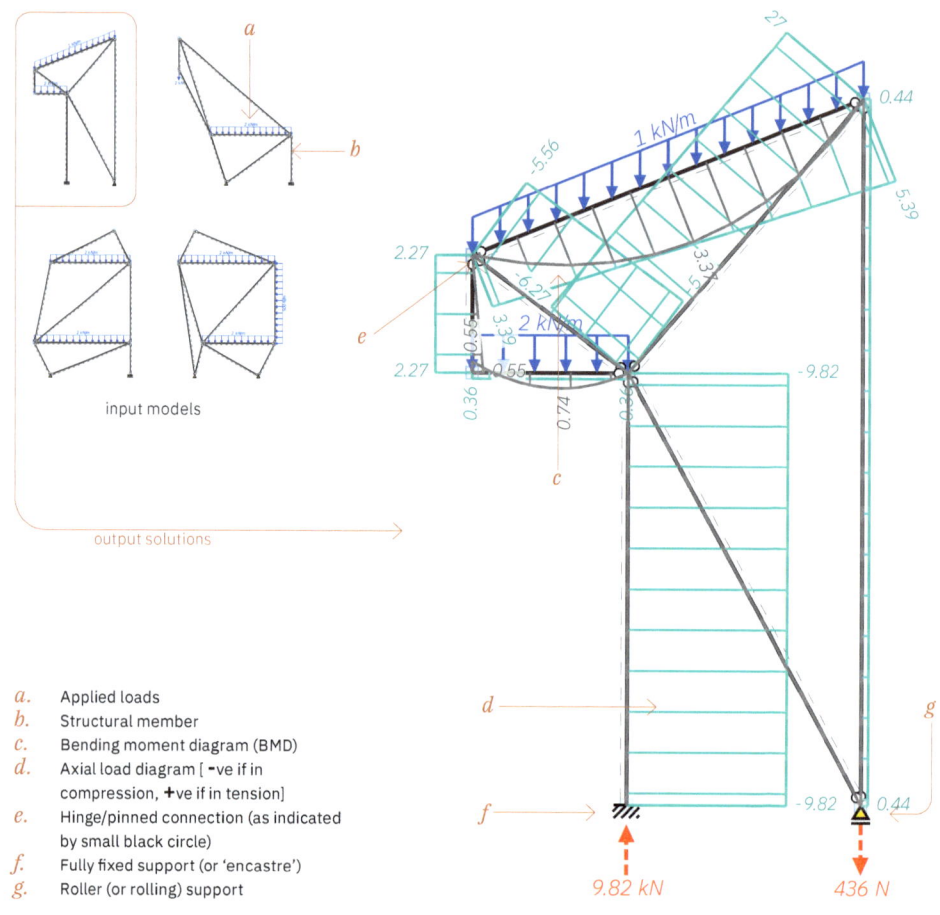

a. Applied loads
b. Structural member
c. Bending moment diagram (BMD)
d. Axial load diagram [ -ve if in compression, +ve if in tension]
e. Hinge/pinned connection (as indicated by small black circle)
f. Fully fixed support (or 'encastre')
g. Roller (or rolling) support

The 'STRIAN' web-app site allows the diagrammatic creation and solution of two-dimensional structural arrangements —Statics, this process reveals some of the force implications for a proposed frame. This in-turn can help to determine where reinforcement or alternative decisions are required. Keep in mind that most buildings require a three-dimensional resolution of statics, thus discuss the application of this method with a structural engineer when using this method.   http://structural-analyser.com/

*fig.7_4: [Above left 4] 2D Structural analysis web-app, four different models entered*
*fig.7_5: [Above right and following 3] 2D Structural analysis web-app, four 'solutions'*

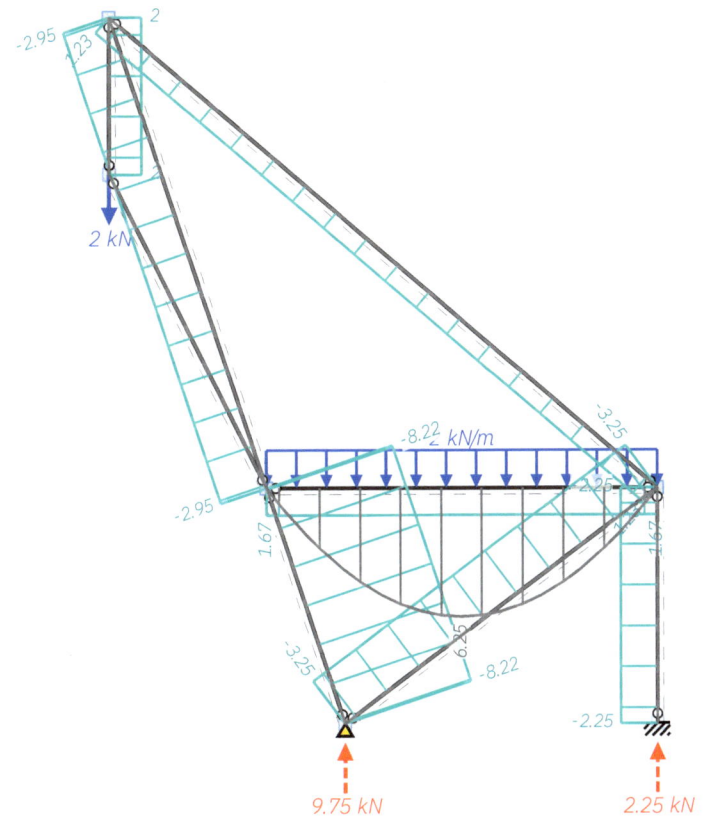

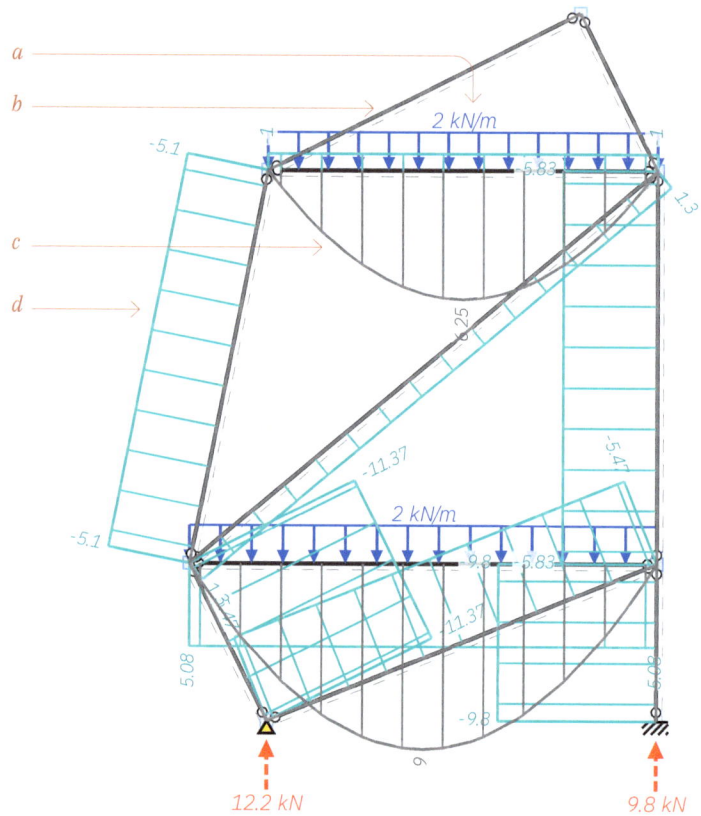

a. Applied load. In this case the load is a uniformly distributed load (UDL) of 2kN/m.
b. Structural member
c. The Bending Moment Diagram (BMD) is a graph showing the bending, or flexural, forces acting in a member. By looking at the BMD one can see whereabouts in the member the flexural forces are greatest. Here a UDL of 2kN/m results in a curved BMD, with the greatest bending moment occurring at mid-span.
d. The width of the axial load diagram shows the magnitude of the axial force acting in the member. Here the axial load in the member is -5.1kN. Note the sign convention: a negative number indicates compression; a positive number would indicate tension.

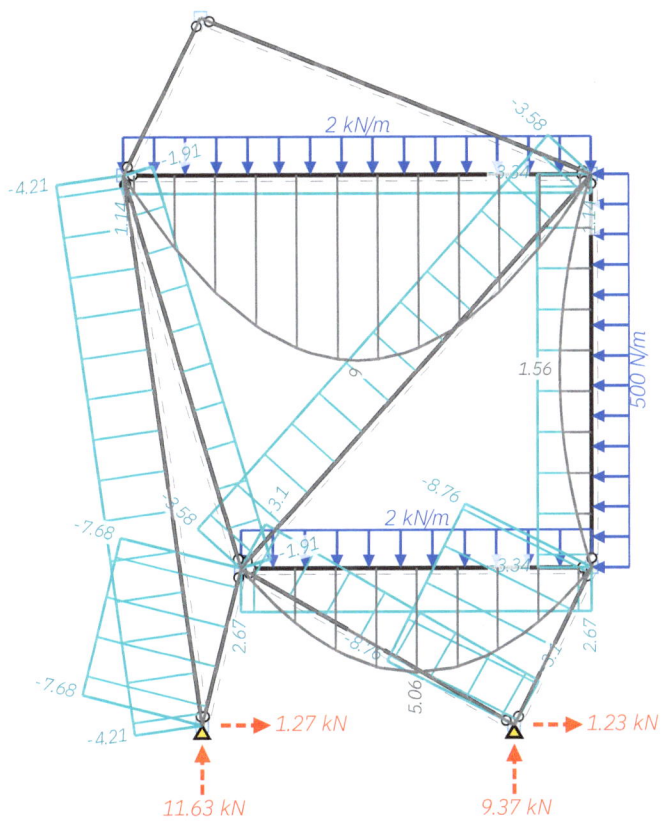

fig.7_6: [Opposite and above] 2D Structural analysis web-app 'solutions'

Web Applications

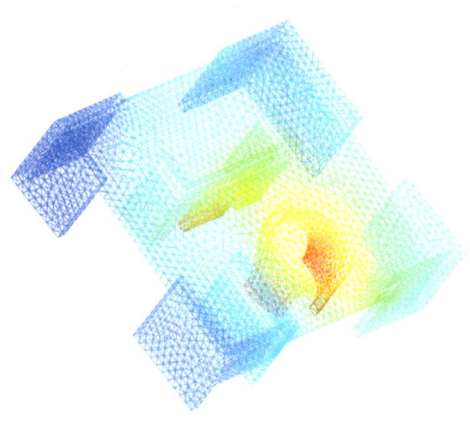

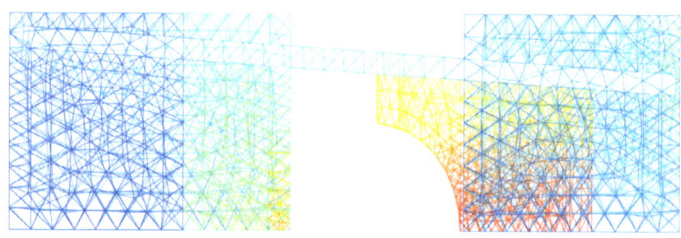

The 'Simscale' Web-app allows for Computational Fluid Dynamics (CFD) analysis. This example (above) is an analysis of a Thermo-structural Heat Transfer model. We can see how the heat dissipates from select points (in red) through the remainder of building—in this instance it is made of one solid material. Full documentation online, requires a little practice. Here the test model was made using Onshape.com.   http://www.simscale.com

*fig.7_7: 2D CFD web-app: Thermo-structural Heat Transfer simulation. Axonometric and profile views*

*fig.7_8: [Opposite] 2D CFD web-app: Thermo-structural Heat Transfer simulation. Top view*

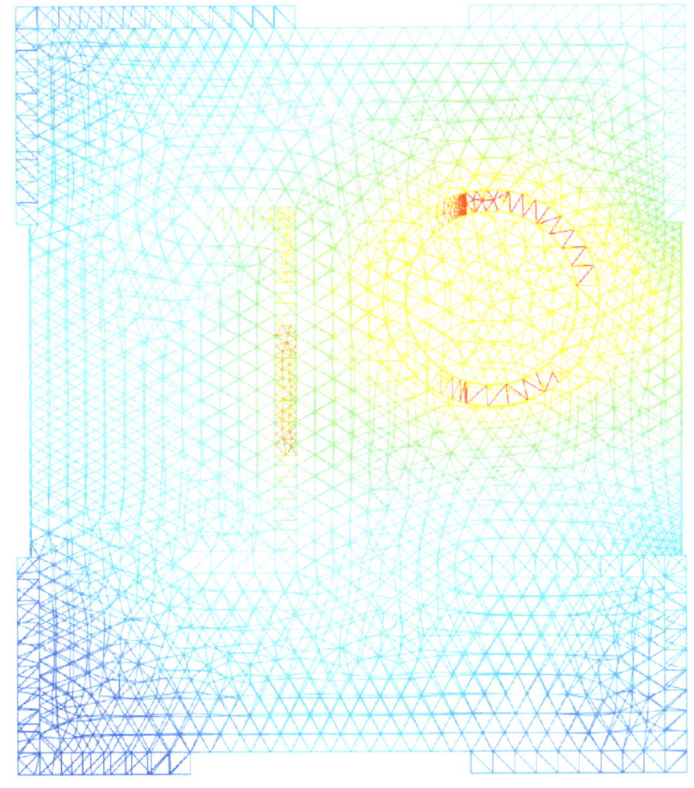

Web Applications

This second example is a wind turbulence analysis of an urban scale model. Here we can see how a particular wind speed vector behaves after passing the modelled tower, and how it might interact with the variation of ground level buildings. Full documentation online, requires a little practice.

The following represent the key parameters for the above simulation: Mesh [ Algorithm: Hex-dominant, Mode: External, Sizing: Manual, Min Edge: 1m ]. Simulation [ Analysis: Incompressible, Turbulence: k-omega SST / Materials: Air / Boundary Conditions: Velocity Inlet (select incoming mesh face) ~<25m/s, Pressure Outlet: (select outgoing mesh face) 0 Pa, Wall: 'No-slip' + select building elements, Wall: 'Slip' + select enclosing mesh facets ]. Simulation Run: Solution Field: [ Results: SCL Pressure, SCL Velocity, Particle Traces >> Settings >> Compute Vector: Velocity ] + Seeds >> Spacing: ~ 15m ]

Here (as for Fig.7_3) the test model was made using Onshape.com. http://www.simscale.com

*fig.7_9: [Above] 2D CFD web-app: Incompressible Turbulence simulation. Axonometric view*

*fig.7_10: [Opposite] 2D CFD web-app: Incompressible Turbulence simulation. Profile and Top views*

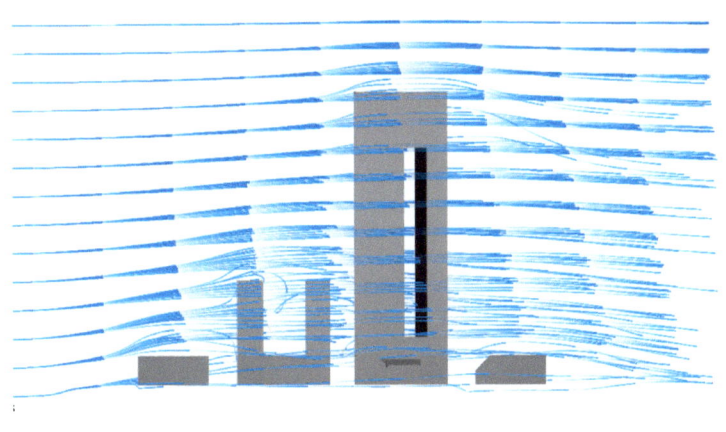

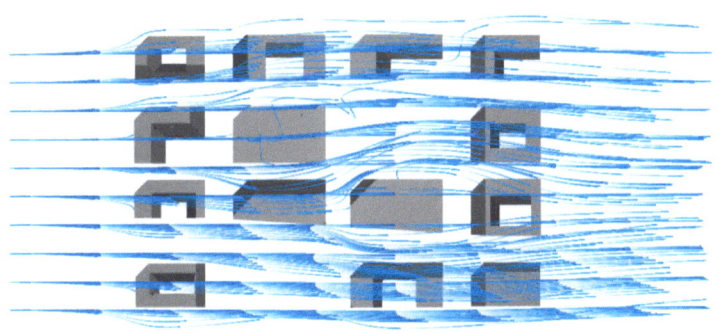

# The Floating University, Berlin

Raumlabor and Foerster-Baldenius apply an altogether different attitude to construction, combining scaffold component and reappropriated systems with inflatables. The project itself is taken as a prototype and sought to provide a educational context for discussing architectural futures. We see here its water filtration process, designed by Katherine Bell the artist in residence, using plants, mushrooms, biofilms, sand, activated carbon, molluscs, and bacteria stored across nine spiralling connected cascading bathtubs. A manually operated wheel is used to draw basin water from the ditch through the tubs, themselves part of a wider water handling system that also manages rainwater, greywater and blackwater whilst supplying all water needs, feeding a greenhouse of 35 different varieties of tomatoes, and supplying methane gas for cooking.

*fig.7_11: Cascading bathtub grey water filtration system*

*fig.7_12: [Opposite] East, South and North Elevations*

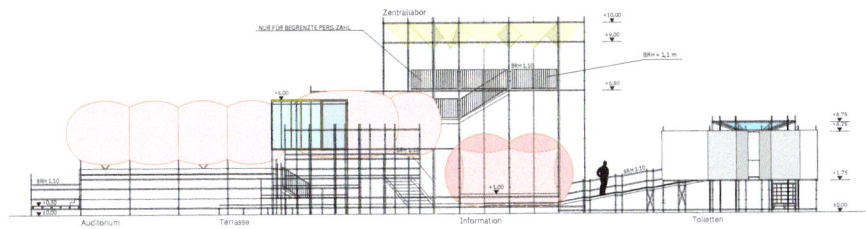

Ansicht 1 / Ostfassade

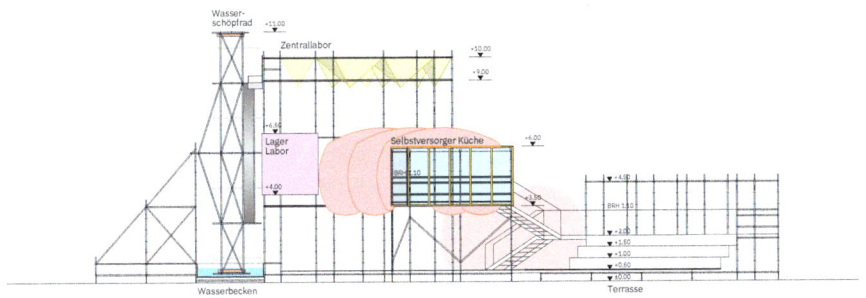

Ansicht 2 / Südfassade

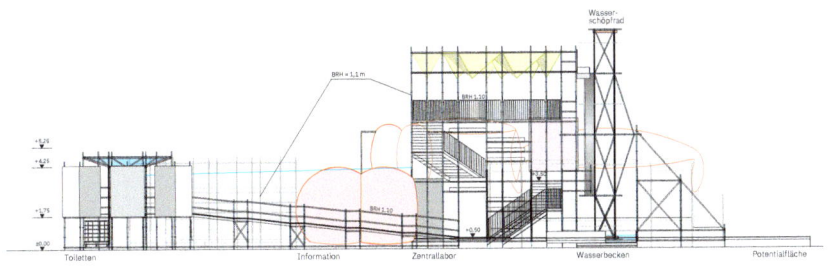

Ansicht 3 / Nordfassade

fig. 7_13: View across basin

fig. 7_14: [Opposite] Scaffold framed inflated spaces

fig. 7_15: [Opposite below] Kitchen area and cascading bathtub grey-water filtration system

## The Eden Project, Cornwall

Designed by Grimshaw & Partners, engineered by Anthony Hunt Associates and built in 2001, the Eden Project is a large geodesic structure consisting of two biomes that house a variety of plants, as a sophisticated greenhouse these environments simulate temperate and subtropical conditions. The lightweight steel structure contains integrated hexagonal cells made of inflated pillows of ETFE (Ethyl Tetra Fluoro Ethylene) to form continuous integrated intersecting shells spanning over 110m. The air contained within the cells provides transparent insulation, permitting sunlight to heat the interior whilst then containing the heat. In addition, on cooler days the cell air pressure can be increased for more insulation, or decreased on hotter days to allow more heat to escape.

Amongst other examples ETFE was also used in the facade of the Beijing National Aquatics Center by PTW Architects and Arup Engineers. For the Eden Project the total surface area is over 39 500 m² and although each ETFE cell weighs less than 1% of its equivalent area in glass the structure is strong enough to support the weight of a small car. As a project it echoes many of the tectonic aspirations of both Archigram and Buckminster Fuller.

The accompanying detail combines several concerns—drainage, pneumatics, waterproofing, structure, and ETFE framework itself containing removable fixings (for maintenance)—all coordinated into the most compact possible profile.

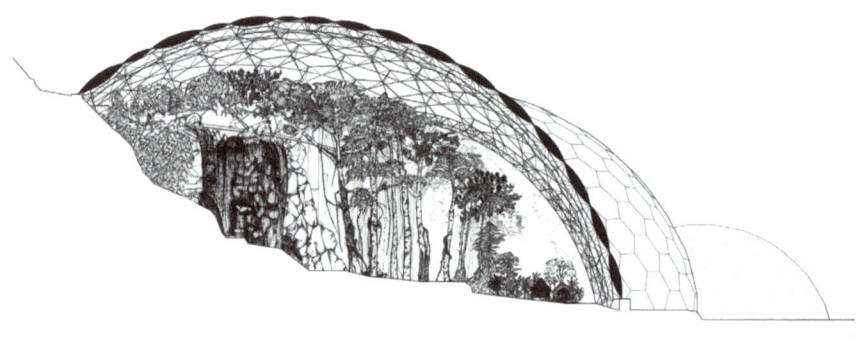

fig.7_16: Eden Project. Cross-section
fig.7_17: [Right] The Eden Project
fig.7_18: [Opposite] Roof Detail

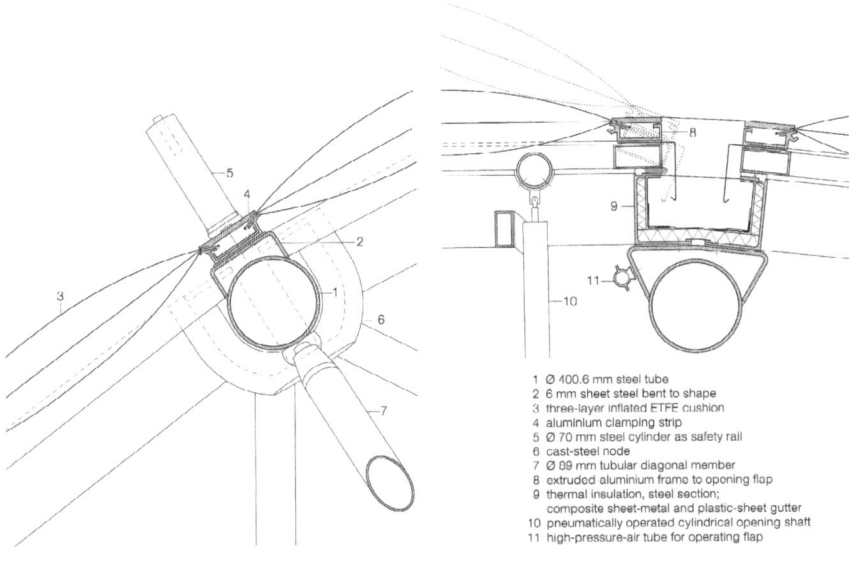

1. Ø 400.6 mm steel tube
2. 6 mm sheet steel bent to shape
3. three-layer inflated ETFE cushion
4. aluminium clamping strip
5. Ø 70 mm steel cylinder as safety rail
6. cast-steel node
7. Ø 89 mm tubular diagonal member
8. extruded aluminium frame to opening flap
9. thermal insulation, steel section; composite sheet-metal and plastic-sheet gutter
10. pneumatically operated cylindrical opening shaft
11. high-pressure-air tube for operating flap

## The Crystal Palace, Hyde Park

Designed by Joseph Paxton to house the Great Exhibition of 1851. This was an early predecessor to modular strategies similar to those employed in the Eden Project. The building was composed of plate glass, iron and wood units and enclosed a 92 000m² space of some 14 000 exhibitors showcasing discoveries and current technology of the industrial revolution. The building was moved to Sydenham South London a year later before burning down in 1936 after rapid fire spread across the wooden flooring within.

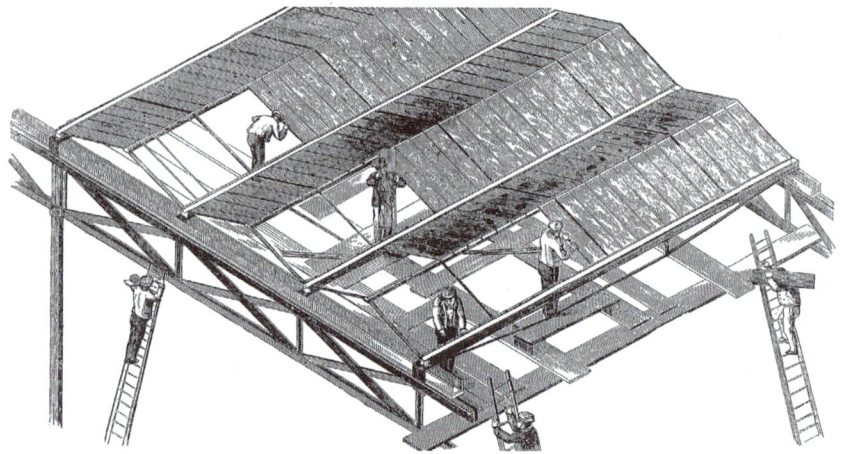

*fig.7_19: Axonometric Detail, Crystal Palace build process. Component and modular hierarchy is echoed in Grimshaw & Partners Eden project*

*fig.7_20: Queen Victoria opens the Great Exhibition in The Crystal Palace in Hyde Park, London, in 1851*

## The London Zoo Aviary, 1965, Lord Snowdon, Cedric Price and Frank Newby.

Unlike the large spanning repetitive modular strategies employed in projects like The Crystal Palace and the Eden Project but still with an environmental focus, the London Zoo Aviary uses a tensile structure wrapped with an permeable mesh. It was one of the first projects to use aluminium as a lightweight structural material.

fig.7_21: The London Zoo Aviary

## Prada Transformer, Seoul

Designed by the Office of Metropolitan Architecture (OMA) the PRADA Transformer was a rotating temporary event space designed to accommodate a different type of activity on each of its internal faces. The design strategy shares the same framework and skin principle as the Snowdon Aviary, but utilises a necessarily more robust steel framework and more experimental material as skin. The steel framework was designed to distribute vertical loads efficiently in any rotation, whilst also moveable via focussed crane connection points. The cladding material was originally designed as a protective wrapper for large pieces of industrial machinery placed in storage for long durations, this project marked its first application as architectural cladding.

*fig.7_22: Rotation in Progress*

*fig.7_23: [Below] Event positions*

*fig.7_24: [Opposite] Interior: Fashion Exhibition rotation*

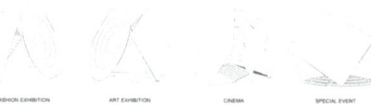

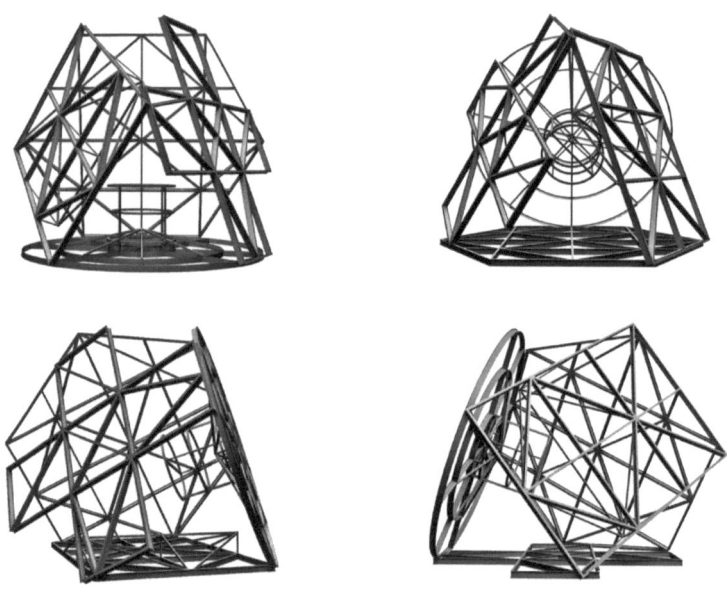

fig.7_25: Steel Frame (Four positions)

fig.7_26: [Opposite] Pad Foundation Arrangement

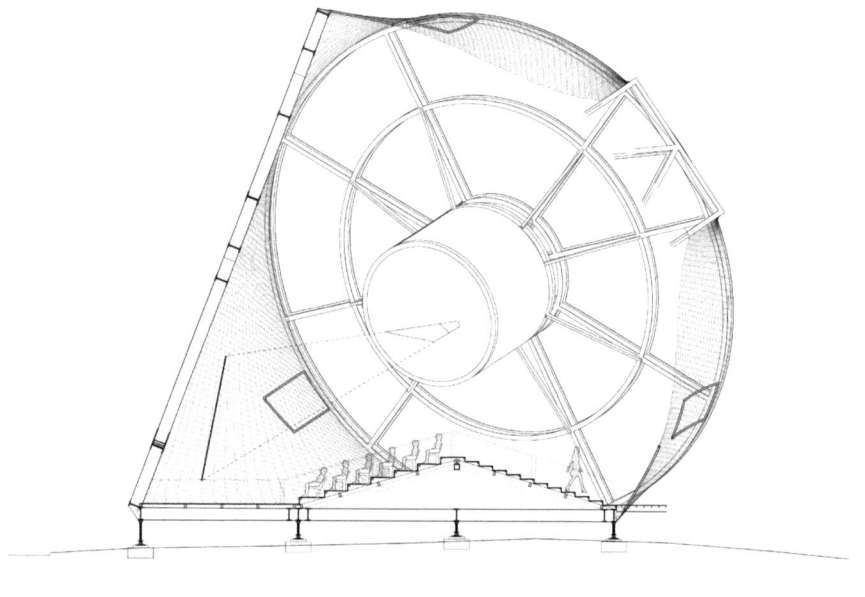

CINEMA

fig.7_27: Section through Cinema rotation
fig.7_28: [Opposite] Plan states

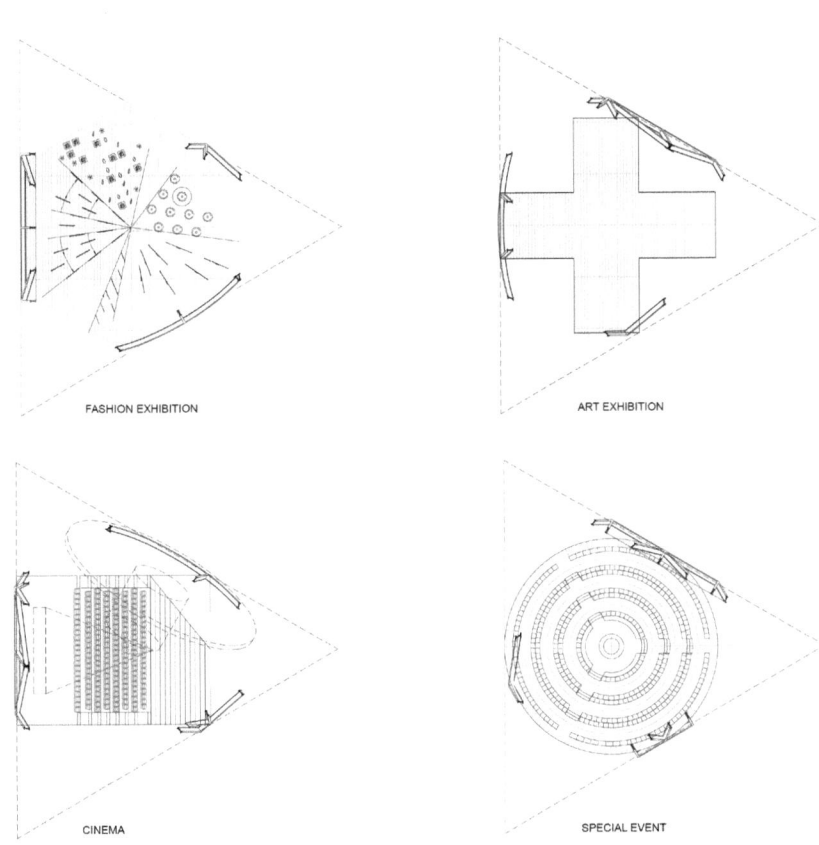

## Paul Smith Shop, London

This final reference project, the Paul Smith shop by 6a Architects, addresses precision and coordination—further aspects of detail design. The facade window jamb detail featured here reflects an alignment of consultants, fabricators, local references, materials and techniques combined. The drawings by facade consultants Montrésor Partnership detail the meeting of elements and speak something of the roles played by each: where the cast iron cladding serves as a rain-screen defining a pressure equalised vented and drained cavity behind. Beyond, the waterproof membrane and thermal insulation are fixed to the reinforced concrete primary structure. Panel weight and positions are supported by a system of horizontal stainless steel rails and vertical channels, these carry the loads

back to the concrete without compromising the waterproof or thermal layers - a build up not uncommon for cladding on concrete structures. Adjacent to this we see the curved glass of the shop display, echoing the nearby glazing of Burlington Arcade, supplied by a glazing manufacturer engaging car windscreen technology, and supported on a CNC (Computer Numerical Control) cut metal carrier frame - supplied by an additional fabricator. In a parallel, digital processes were used in the reinterpretation of a nearby balustrade pattern to generate a continuous relief geometry, then translated to polyurethane moulds through CNC milling for the sand casting process, before eventually casting the iron panels.

As such, we see the matter of manufacturer tolerance; of the curvature of the glass meeting the framework, then meeting the cladding; through the mediation of casting, welding, cutting, adhesion and fixing all require communication - further enabled through drawing. Here we see a merger of traditional and digital processes in both design and production.

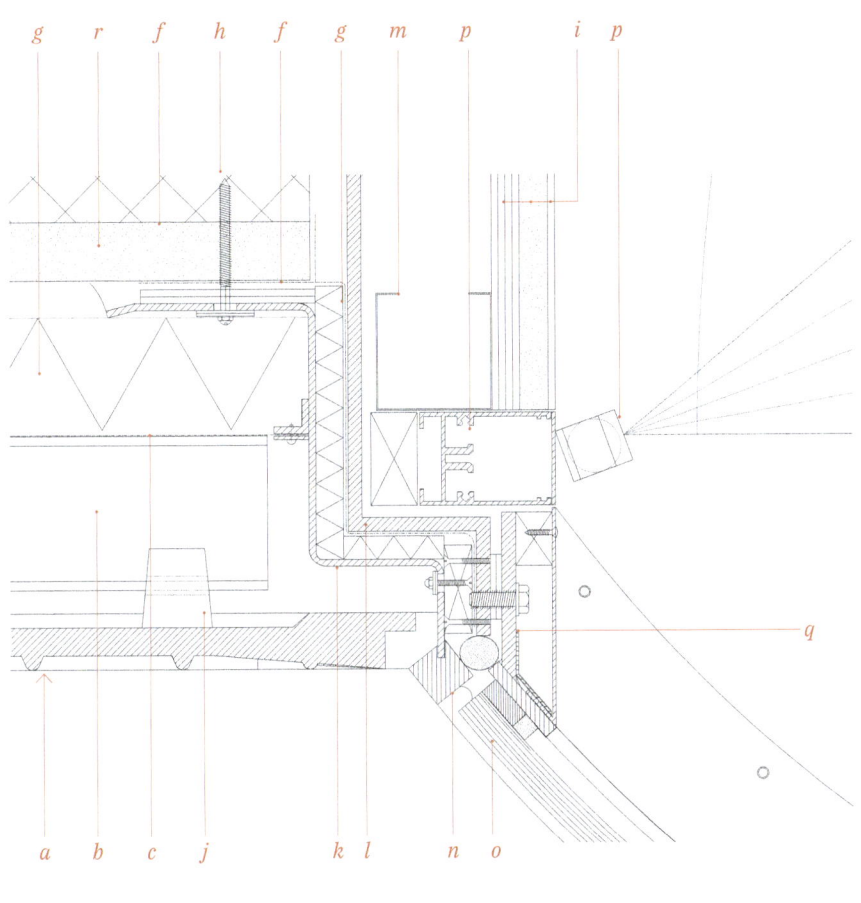

- *a.* Cast-iron panel (Cladding)
- *b.* Steel horizontal support bracket
- *c.* Breather membrane
- *f.* Waterproof membrane
- *g.* Insulation
- *h.* Concrete primary structure
- *i.* Interior: Plywood, Plasterboard, Skim plaster lining.
- *j.* Cast-in panel hook
- *k.* Pressed Brass shroud
- *l.* Galvanised steel plate
- *m.* Metal drylining stud
- *n.* Brass frame
- *o.* Curved laminated glazing
- *p.* Concealed lighting track, and display lighting.
- *q.* S/steel glazing carrier frame
- *r.* Retained existing render

*fig.7_29: [opposite] Paul Smith facade cast iron panelling and projecting curved windows*

*fig.7_30: Architect detail plan drawing of typical window jamb, a convergence of elements*

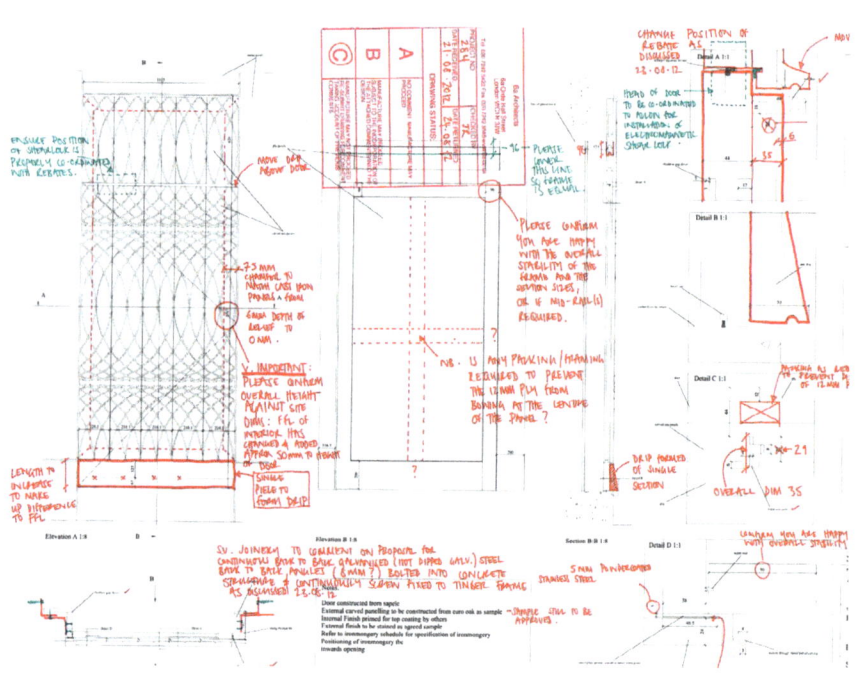

fig.7_31: Cast-iron facade cladding panel (a.) fabricators drawings with architect comments
fig.7_32: [Opposite] CNC Stainless steel glazing carrier framework (q.) fabricators drawings with architect comments

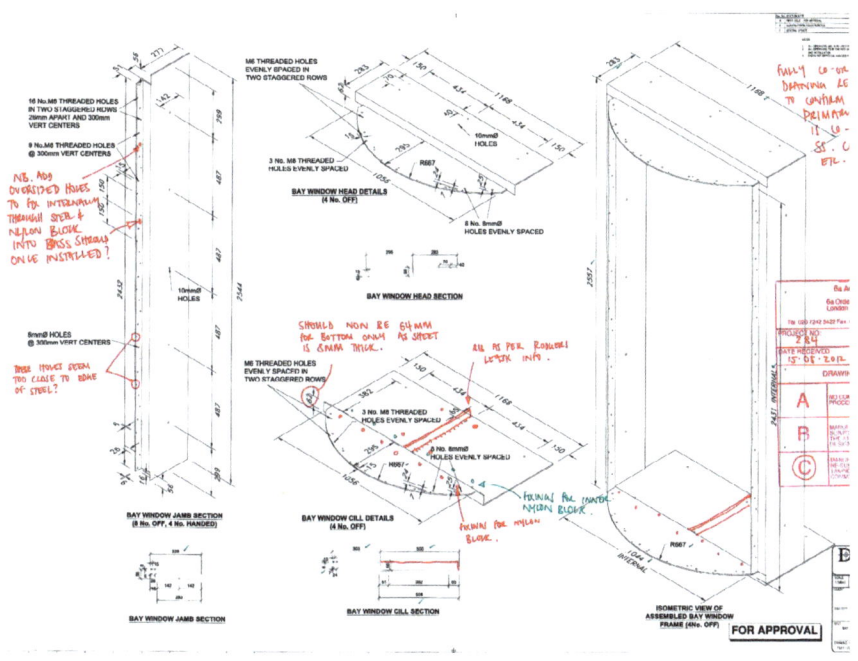

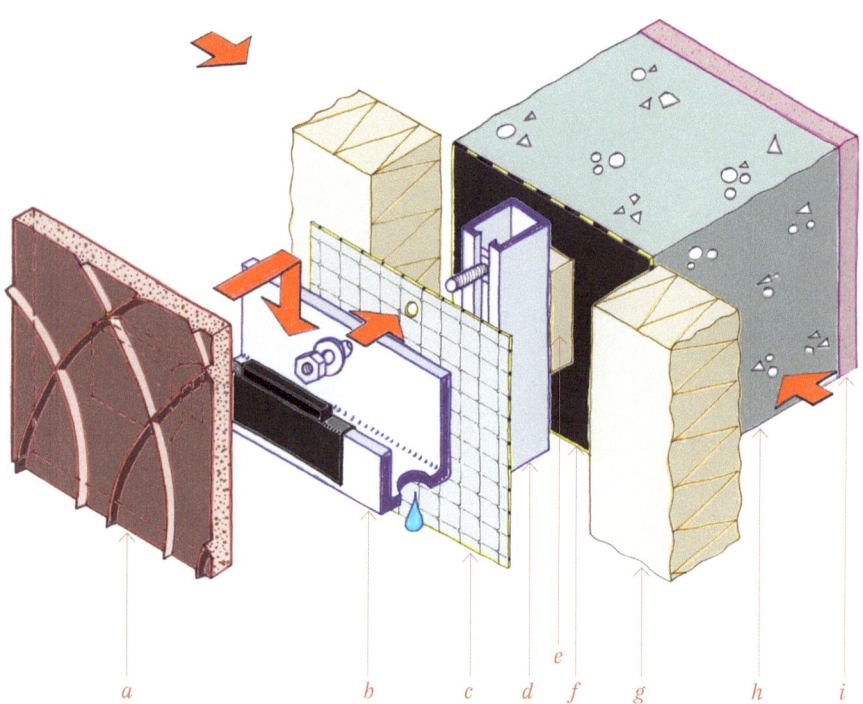

*a.*    Cast-iron panel (Cladding)
*b.*    Steel horizontal cladding support bracket with drainage openings
*c.*    Breather membrane
*d.*    Steel vertical Halfern support channels for horizontal cladding brackets (matching adjacent insulation depth)
*e.*    Intermittent insulation providing a thermal break (maintaining the insulation line)
*f.*    Waterproof membrane
*g.*    Insulation
*h.*    Concrete primary structure
*i.*    Interior lining: Plywood, Plasterboard, Skim finish

*fig.7_33: [Opposite] Facade consultant drawing illustrating cladding sub-structure and build-up*
*fig.7_34: [Top left] local balustrade geometry*
*fig.7_35: [Top right] CNC application of new geometry to polyurethane*
*fig.7_36: [Bottom left] sand cast iron panelling*
*fig.7_38: [Bottom right] cast-iron rainscreen cladding installed*

Ending / beginning

The descriptions of the preceding projects—varying broadly in scope, time, and approach—have only lightly touched upon the surface of what is possible. In exploring your own technical interests, as indicated in the earlier 'Case-study' section of Chapter 1 and 5d, relevant projects will need to be studied and discussed in much greater depth than was possible in the examples mentioned.

Each has come to embody an approach, whether by an individual or an office, towards the technical realisation of an idea. When conducting your own research, short informal first hand conversations with (where possible or convenient for all parties) the practitioners involved is also encouraged, not least as a means to gain a fuller picture of a project's story.

Together, these projects reflect a small subset of a broad spectrum of technical explorations.

Further projects are highlighted throughout the books of the reading list of Chapter 2: The Field, the technical knowledge imparted therein will provide the skills needed to recognise and dissect projects as the TS research unfolds. Where combined with experimentation, discussion, and prototyping, the basis of a three-dimensional technical exploration should emerge.

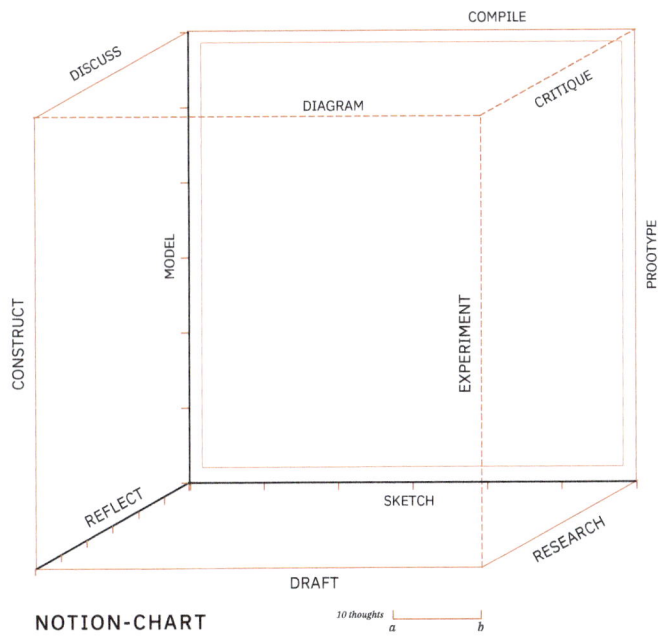

fig.7_38: Notion-chart. A TS themed interpretation of the 'Ocean Chart' illustration to chapter 2 of Lewis Carroll's "The Hunting of the Snark", 1876

*fig.8_1: Circle maze*

*fig.8_2: Square maze*

*Notes*

*East Winter Books*

www.ingramcontent.com/pod-product-compliance
Lightning Source LLC
Chambersburg PA
CBHW041508010526
44118CB00006B/183